Nacera LEKHAL

The benefits of stand-alone energy systems

Nacera LEKHAL

The benefits of stand-alone energy systems

Renewable energies

ScienciaScripts

Imprint
Any brand names and product names mentioned in this book are subject to trademark, brand or patent protection and are trademarks or registered trademarks of their respective holders. The use of brand names, product names, common names, trade names, product descriptions etc. even without a particular marking in this work is in no way to be construed to mean that such names may be regarded as unrestricted in respect of trademark and brand protection legislation and could thus be used by anyone.

Cover image: www.ingimage.com

This book is a translation from the original published under ISBN 978-620-6-70481-2.

Publisher:
Sciencia Scripts
is a trademark of
Dodo Books Indian Ocean Ltd. and OmniScriptum S.R.L publishing group

120 High Road, East Finchley, London, N2 9ED, United Kingdom
Str. Armeneasca 28/1, office 1, Chisinau MD-2012, Republic of Moldova, Europe
Printed at: see last page
ISBN: 978-620-7-85960-3

Copyright © Nacera LEKHAL
Copyright © 2024 Dodo Books Indian Ocean Ltd. and OmniScriptum S.R.L publishing group

Contents

Preface

Climate change is one of the greatest challenges of our time. It is, however, equally important to ensure access to energy in order to promote quality of life and economic development. It is therefore essential to address this issue as part of the sustainable development agenda. Current progress in the development of new technologies has given confidence and hope that energy targets will be met. Drastic cost reductions and technological advances in wind turbines and solar photovoltaics have shown that renewable energy resources can play an important role in the world's electricity systems and that the long-awaited breakthroughs in storage technologies are set to change the energy mix significantly. [2]

These developments led to the assumption that we were 'done' with fossil fuels in the energy system, that there was no longer any need to develop new resources and that we should stop using them as quickly as possible. This assumption also gave the impression that there were "good" technologies, based on renewable energies, on the one hand, and "bad" technologies, based on fossil fuels, on the other. In fact, this debate is much more nuanced and requires in-depth examination. Carbon capture and storage (CCS) techniques and the management of methane emissions throughout the fossil fuel value chain can make it possible to achieve the ambitious targets for reducing CO_2 emissions while fossil fuels remain part of the energy system. In this way, fossil fuels would become "part of the solution" rather than "part of the problem". All technologies have a role to play in an energy system driven by the search for savings. [2]

Today, 74% of the world's energy is produced from fossil fuels (oil, coal and gas), 20% from renewable energies (hydro, biomass, solar, wind) and 6% from nuclear power. Numerous studies on the depletion of fossil fuels point to the following conclusion: the quantity of fossil fuels available will decrease between 2010 and 2020 and will be exhausted before the end of this century. Our energy future must be based on nuclear and renewable energies.

Current nuclear power generation offers very high power density and environmental benefits in terms of CO2 emissions. However, this form of energy has a number of drawbacks, including the difficulty of reprocessing waste and buildings, its impact on the environment, safety problems and the fact that its fuel is non-renewable (depletion of uranium 235 is estimated at the end of the century). Despite extensive research to solve the problem of waste and to develop new generations of fast-breeder reactors with larger fuel reserves, the average level of safety and the human and ecological consequences of a nuclear accident remain the major drawbacks of this technology. Although it is difficult to imagine eliminating this energy solution, it is preferable to limit it to its lowest level of necessity. [1]

In other words, if current trends continue, if the current share of fossil fuels remains unchanged and if energy demand doubles between now and 2050, emissions will far exceed the amount of carbon that can be emitted if we want to limit the rise in average temperatures to 2°C. This level of emissions will have disastrous consequences for the planet's climate. We have the opportunity to reduce emissions in the energy sector, in particular by reducing energy consumption as well as the net carbon intensity of the ënergëtic sector by switching fuels and controlling ë CO_2 emissions. [2]

Reducing emissions does not mean ruling out the use of fossil fuels, but significant change is imperative; business as usual will not reduce them. Energy efficiency and renewable energies are often seen as the only solutions needed to achieve climate targets in the energy system, but they are not enough. Increasing the use of CSS techniques will be essential, which should enable an annual reduction of 16% by 2050. This assertion is based on the Fifth Assessment Report of the Intergovernmental Panel on Climate Change, which estimates that limiting emissions from the energy sector without CSS techniques would increase the costs of mitigating the effects of climate change by 138%. [2]

Renewable energies cannot be uniformly used in the energy system to replace the use of fossil fuels, in particular because of differences in the capacity of the energy sub-sectors to switch from fossil fuels to renewable energies. For example, in some industrial applications such as cement and steel production, emissions come from both energy use and production processes. As alternative technologies likely to replace current production techniques are not yet available in the foreseeable future, these techniques are likely to continue to be used in the short to medium term. In some cases, CSS techniques can provide a solution compatible with current demand and give the time needed to develop other solutions. [2]

There are many isolated sites in the world powered by autonomous electricity generation systems. These generators use iocal renewable sources. They include photovoltaic panels, wind turbines and microturbines. Electricity from renewable sources is intermittent and dependent on climatic conditions. These renewable generators are coupled to a storage system to ensure continuous availability of energy. [1]

The development of hydrogen technologies has been very significant over the last ten years. The progress made means that the performance of the "hydrogen storage system", a gas storage unit and a fuel cell, can be expected to be excellent. However, the performance of the hydrogen storage system has not been reviewed. Furthermore, the day-to-day use of this storage system to increase heat generation has never been discussed. [1]

The development of integrated energy networks, with common operating regimes, provides an excellent opportunity to strengthen the links between technologies, encouraging the penetration of less carbon-intensive and economically efficient technologies. Whether we like it or not, fossil fuels will be part of the energy system for decades to come. They will continue to underpin social and economic development around the world. [2]

Chapter pedagogy

Each chapter begins with a one-page introduction that includes a list of the sections.

Each chapter section begins with a brief introduction to the subject.

A number of examples help to clarify and illustrate particular concepts or procedures.

At the end of each chapter is a summary. At the end of the book.

In the first chapter, we describe all the devices used to produce electrical energy, the concepts of energy transformation, and non-renewable energy sources (fossil and nuclear). Renewable energy sources.

In the second chapter we will discuss wind energy, its history, principle and structure, characteristics and sizing, map of wind power in Algeria, wind farms and power, standards, advantages and disadvantages. Example of a wind farm.

In the third chapter, we will present the Hybrid Systems (Hydrolienne, Principe de fonctionnement de l'hydrolienne, Les différents types d'hydroliennes et les exploitants,...).

In the fourth chapter, we look at photovoltaic solar energy, the principle of a photovoltaic installation, solar resources in Algeria, photovoltaic cell technologies, photovoltaic modules, MPPT, photovoltaic characteristics and connectors, standards. The inverter (role, principle, characteristics and efficiency). Example of a photovoltaic installation.

And finally, in the fifth chapter, we look at other sources of renewable energy, the families of renewable energies (solar energy, wind energy, hydraulic energy, biomass, geothermal energy). The different types of renewable energy in the world. Profitability.

This book is intended for the **Electronics** course, the aim of which is: to arouse the student's interest in renewable energies in general and in energy systems using solar or wind power in particular. To enable the student to acquire a degree of competence in the dimensioning of a wind or photovoltaic installation.

We have tried to base ourselves on various basic works as mentioned on the reference page.

CHAPTER 1

Electric power generation systems

Transformed energies (solar energy, wind energy, geothermal energy), as opposed to natural energies (new or renewable), are created by one or more transformations from a natural energy source.

Overview of the Chapter: Electric power generation equipment

1.1 Notions on energy transformations

1.2 methods of producing electrical energy

1.3 Non-renewable energy sources

1.4 Renewable energy sources.

1.1 Notions on energy transformations [3]

When we want to use energy, we can't do it in its primary form. Special cases: drying and solar heating. It has to be transformed. Transformation requires the use of more or less sophisticated processes and technologies.

A simple match to burn wood

A thermal power station to produce electricity

An engine to produce mechanical energy ...

1.2 Energy end-use [3]

The end uses of energy are :

Thermal Energy

The higher the temperature required, the more complex the transformation process and technology, and the higher the price.

Low temperature (30 to 120°C): domestic hot water, space heating, absorption machines, etc.

Medium temperature (100 to 500°C): drying, cooking, sterilisation, distillation,

High temperature (500 to 1800°C): Glassworks, cement works, metallurgy, chemical treatments,

Luminous Energy

It is obtained from heat. The higher the level of illumination, the more complex and expensive the process and technology.

Mechanical Energy

In transport, the higher the speed required, the more complex and expensive the process and technology, and the higher the energy consumption.

Thermal engines (cars, mopeds, motorbikes, coaches, buses, trains, planes, boats, tractors, steam engines, internal combustion engines (diesel, internal combustion), turbines, ...) are used in the production of electricity.

Electric motors (cars, trams, trains, electrical appliances, automation, industry, etc.)

Information Energy

It is currently expanding rapidly.

It is generally obtained from electricity using different technologies:
Television, Fax, Telephone and telecommunications, Computers.

1.3 Energy classification

Energy sources are mainly of fossil origin (oil, gas, coal, etc.). Blocked chemical energy can be released by oxidation reaction (combustion). The clean energy available (thermal, electrical, kinetic) is negligible.

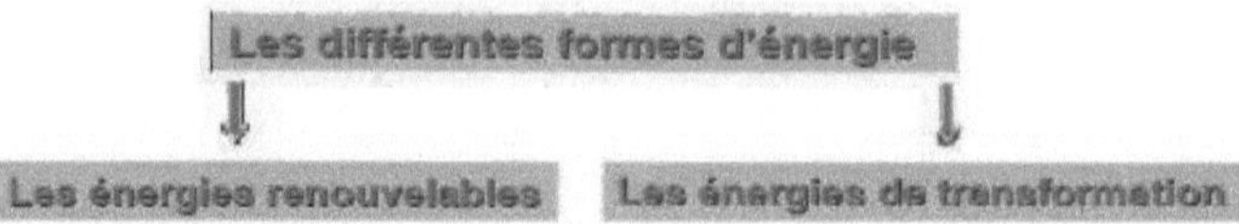

Figure1.1: energy classification

Energy for transformation

The three basic energies are

- Thermal
- Electric
- Mecanics

1.2 Electricity generation methods [3]

Energy conversions can take place in succession, with an accumulation of energy losses.

The most significant and classic example relates to the production of high-power electrical energy (thermal or nuclear power stations).

The thermodynamic cycle comprises the following elements:

- Boiler (thermal energy)
- Turbine (mechanical power)
- Condenser

The turbine drives an alternator producing electrical energy which, after distribution, can be used to produce mechanical energy (fan) or thermal energy (heating).

The energy chain is shown in Figure (1.2, 1.3):

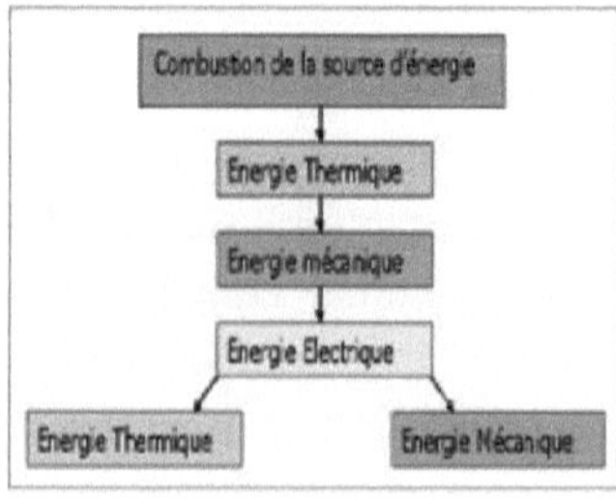

Figure 1.2: The energy chain

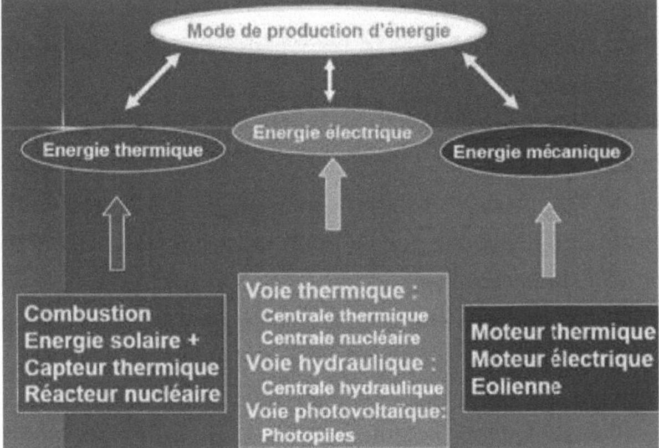

Figure1.3: Energy production method

1.3 Non-renewable energy sources

1. Fossil fuels

Fossil fuels (oil, natural gas and coal) are the raw material of the chemical industry and the most widely used source of energy in the world: they provide more than 80% of the energy used, far ahead of nuclear energy and other forms of energy (hydro, wind, solar, etc.). The world's energy needs have increased considerably over the twentieth century, and the development of emerging countries such as China means that we can expect an even more rapid increase over the next few decades. The International Energy Agency predicts that demand over the next twenty-five years will require production equal to that of one hundred and fifty years of fossil fuel exploitation. But resources are not inexhaustible: these products are formed by a succession of biological and geological mechanisms that take millions of years to complete, so these resources are not renewable on a human timescale. [2]

Benefits [3]

- Practical use
- Rapid energy release
- Great availability
- Mastered storage technology
- Diversification of applications

Disadvantages [3]

- Limited reserves
- Air pollution :
- CO2 and CH4: greenhouse effect

-    Imbalances in HC, CO and CH4 generate ozone on the earth's surface and cause respiratory problems.

-    The sulphur contained in fuel is responsible for the corrosion of thermal installations and acid rain.

Lead-based petrol additives have harmful effects on the nervous system.

2. Nuclear fuels [5]

Nuclear fuel is the product which, containing fissile materials (uranium, plutonium, etc.), supplies energy to the core of a nuclear reactor by sustaining the nuclear fission chain reaction.

The terms **fuel** and **combustion** are totally inappropriate to characterise both the product and its action. In fact, combustion is a chemical oxidation-reduction reaction (exchange of electrons) whereas the "combustion" of radioactive materials comes from nuclear reactions (fission of atomic nuclei). These terms are used by analogy with the heat released by a burning material.

Fissile materials are used for the nuclear propulsion of naval vessels (in particular aircraft carriers) and nuclear submarines, and as fuel in nuclear power plants: a 1,300 MWe pressurised water reactor contains around 100 tonnes of fuel that is changed periodically.

UOX (Uranium Oxide) fuel is made up of uranium dioxide (UO_2) pellets. These pellets are stacked in zirconium alloy tubes. These tubes, which are around 4 metres long, are also known as cladding. The pellets together form a rod. The rods are sealed at both ends and pressurised with helium.

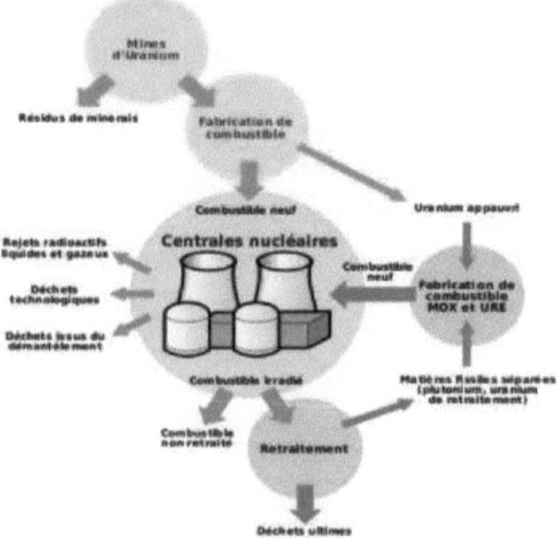

Figure1.4: Nuclear fuels

Benefits [3]

-    Decentralisation: nuclear power plants can be installed wherever you want,

regardless of deposits or other factors.
- Good yield: natural uranium produces 116,000 kWh/kg of energy

Disadvantages [3]
- Require availability and mastery of cutting-edge technologies
- The problems of nuclear waste have not yet been seriously resolved (dangerous contamination for humans).
- Their use is still limited, and more needs to be done.
- The uranium used in nuclear fission reactions is exhaustible.
- A fearsome aspect of nuclear weapons.

1.4 Renewable energy sources.

Renewable energies are sources of energy that are naturally renewed quickly enough to be considered inexhaustible on a human timescale. They come from cyclical or constant natural phenomena induced by the stars: the Sun mainly for the heat and light it generates, but also the attraction of the Moon (marees) and the heat generated by the Earth (geothermal energy). Their renewable nature depends partly on the rate at which the source is consumed, and partly on the rate at which it is renewed. [4]

The expression "renewable energy" is the short and usual form of the expressions "renewable energy sources" or "energies of renewable origin", which are more correct from a physical point of view. physical point of view. [4]

The share of renewable energies in the world's final energy consumption was estimated at 17.9% in 2018, of which 6.9% was traditional biomass (wood, agricultural waste, etc.) and 11.0% "modern" renewable energies.) and 11.0% of "modern" renewable energies: 4.3% of heat produced by thermal renewable energies (biomass, geothermal, solar), 3.6% of hydroelectricity, 2.1% for other electrical renewables (wind, solar, geothermal, biomass, biogas) and 1% for biofuels; their share of electricity production was estimated at 26.4% in 2018. [4]

Benefits [3]
- On a human timescale, the sun, geothermal energy and the wind are inexhaustible sources of energy.
- New energies are clean and well distributed across the globe.

Disadvantages [3]
- Their use is still limited, and more progress needs to be made.

In order to conserve fossil fuel resources and reduce pollutant emissions, the use of these fuels must be rationalised by
- Controlling and optimising combustion installations - Achieving the most complete combustion possible.

- Reducing consumption

From an energy point of view, renewable energies, for example, should enable us to satisfy our requirements to a large extent and offer the level of comfort of the most advanced countries to the whole of the world's population.

Two conditions must be met:

- Optimising consumption (minimising waste and improving conversion efficiency)
- Clean" energy production.

Energy characterisation [3]

There are two quantitative factors that can be used to make energy choices:

- Theoretical energy production efficiency,
- Real production costs.

Theoretical efficiency of energy production [3]

Whatever the energy production process, the theoretical efficiency is expressed by the ratio :

$$R = \frac{\text{énergie produite}}{\text{énergie utilisée}} \qquad (1.1)$$

$$\frac{\text{energy produced}}{\text{energy used}}$$

Real production costs [3].

The real costs of energy before it is distributed must take account of 3 parameters:

- the cost of fuel (nil for a solar or hydraulic system)
- depreciation of energy-generating plant.
- operating and maintenance costs.

Note: We can always define the partial cost, which takes into account the theoretical efficiency and the cost of the fuel.

The real cost of energy at the level of use involves not only the real cost of production defined above, but also the cost of distribution, which depends on the state of the network, and the cost of use, which depends on the state of the appliances using the energy.

Conclusion [3]

In many industrial processes involving the production or recovery of energy, we have to consider heat transfer.

The most common way of producing heat energy is by combustion, which is why it is so important not only to study heat transfer, but also to describe combustion phenomena and deal with heat exchanger problems.

CHAPTER 2

Wind energy

A wind turbine is a device that converts part of the wind's kinetic energy into mechanical energy available on a transmission shaft, and then into electrical energy via a generator.

Chapter Overview: Wind Energy

2.1 History

2.2 Principle and structure

2.3 Characteristics and dimensions

2.4 Map of wind energy potential in Algeria

2.5 Wind farms and power

2.6 Standards

2.7 Advantages and disadvantages

2.8 Example of a wind farm

2.1 History [6]

Until the 19th century, wind energy was used to provide mechanical work.

The oldest use of wind energy is in sailing: there is evidence to suggest that it was used in the Sea Egee Aegean as far back as the 11th millennium BC (see Navigation in Antiquity). Oceania was probably settled by sailing, for long crossings of hundreds or thousands of kilometres across the open sea.

Around 1600, Europe had 600,000 to 700,000 tonnes of merchant ships; according to a more precise French statistic from around 1786-87, the European fleet had reached 3.4 million tonnes; its volume therefore increased fivefold in two centuries. The wind power used to propel these ships can be estimated at between 150,000 and 230,000 hp, without taking into account the war fleets.

The other main use of this energy was the windmill used by the miller to transform cereals into flour or to crush olives to extract the oil; there were also many windmills used to drain the polders in Holland. Holland. The windmill first appeared on the territory of present-day Afghanistan; it was used in Persia for irrigation from the year 600. According to the historian Fernand Braudel, "The windmill appeared much later than the waterwheel. Yesterday, it was thought to have originated in China; more likely, it came from the highlands of Iran or Tibet. In Iran, windmills were probably in operation as early as the seventh century AD, and probably as late as the ninth century", powered by vertical sails mounted on a wheel that itself moved horizontally (...) The Muslims are thought to have spread these windmills to China and the Mediterranean. Tarragona, on the northern edge of Muslim Spain, is thought to have had windmills since the 10th century.

Fernand Braudel describes the gradual introduction of water and windmills from the eleventh to the thirteenth centuries as the "first mechanical revolution": "these 'primary engines' are undoubtedly of modest power, from 2 to 5 HP for a water wheel, sometimes 5, at most 10 for the wings of a windmill. But in an economy with a low energy supply, they represent a considerable increase in power. The older water mill is far more important than the wind turbine. It does not depend on the irregularities of the wind, but on water, which is by and large less capricious. It is more widespread, because of its age and the multiplicity of rivers...". "The great adventure in the West, in contrast to what happened in China where the mill turned horizontally for centuries, was the transformation of the windmill into a vertical wheel, in the same way as watermills. The engineers say that the modification was brilliant, and the power greatly increased. It was this new type of mill that spread to Chretien. The statutes of Arles record its presence in the twelfth century. At the same time, it was being used in England and Flanders. In the 13th century, the whole of France welcomed it. In the 14th century, it was in Poland and already in Muscovy, because Germany had already passed it on to them".

The windmill, which is more expensive to maintain than the water mill, is more expensive for the same amount of work, particularly for milling. But they had other uses too: in the Netherlands, from the 15th century onwards, and even more so after 1600, the *Wipmolen* played a major role in powering bucket chains that drew water from the ground and discharged it into canals. They were thus one of the tools used in the patient reclamation of the soil in the Netherlands. Another reason why Holland is home to windmills is its location at the centre of the great swathe of permanent westerly winds, from the Atlantic to the Baltic.

At the end of the 18th century, on the eve of the industrial revolution, almost all of mankind's energy needs were met by renewable energies, and wind power played a major role in the energy balance, meeting most of the needs of international transport (sailing) and part of domestic transport (coastal shipping and river navigation), as well as the needs of the food industry (windmills). In an attempt to estimate the breakdown of consumption by energy source, Fernand Braudel puts the share of sailing at just over 1%, compared with over 50% for animal traction, around 25% for wood and 10-15% for water mills. He refuses to put a figure on the share of windmills, for lack of data, but does state that "windmills, which are less numerous than water wheels, can only represent a quarter or a third of the power of disciplined waters". We can therefore estimate the total share of wind power (sail + windmills) at between 3 and 5%.

The advent of the steam engine, followed by the diesel engine, led to the decline of wind power in the 19th century; windmills disappeared, replaced by industrial flour mills. By the mid-twentieth century, wind power was only used for pleasure boating and pumping (agriculture, polders).

Subsequently, for several decades, wind power was also used to generate electricity in remote locations that were not connected to an electricity grid (houses, farms, lighthouses, ships at sea, etc.). Installations without energy storage meant that the need for energy and the presence of wind power had to be simultaneous. The mastery of energy storage using batteries has made it possible to store this energy and use it when there is no wind, although this type of installation only concerns domestic needs and is not applied to industry.

Since the 1990s, improvements in wind turbine technology have made it possible to build wind turbines of more than 5 MW [2]and the development of 10 MW wind turbines is underway. Government subsidies have enabled them to be developed in a large number of countries. Today, these wind turbines are used to produce alternating current for the electricity grids, in the same way as a nuclear reactor. reactor nuclear , a hydroelectric dam or a thermalstationpower thermal . However, the power produced, the production costs and the environmental impact are very different.

Collector structures are becoming increasingly efficient. In addition to the mechanical characteristics of the wind turbine, the efficiency of the conversion of mechanical energy into electrical energy is very important.

2.2 Principle and structure

A wind farm or aerogenerator produces electricity using the force of the wind.

At the top of the wind turbine are the blades, which can measure up to 120m. These blades, also known as rotors, rotate under the effect of a wind of less than Km/h.

The blade system rotates around a hub, the hub of an axle that is retracted into the nacelle, which is carried on the mast and this mast is fixed to the foundations that allow the wind turbine to stand upright.

The nacelle is the main element of the wind turbine, which is automatically oriented towards the wind using sensors to capture the maximum amount of wind, when the system of blades rotates the axis (hub) this speed is not sufficient to produce electricity a multiplier that increases this speed to 1500 rpm and transmits it to a second axis of the generator, The interaction between the rotor's electromagnet and the stator's coil of copper wires produces an electric current, which is then distributed to the grid.

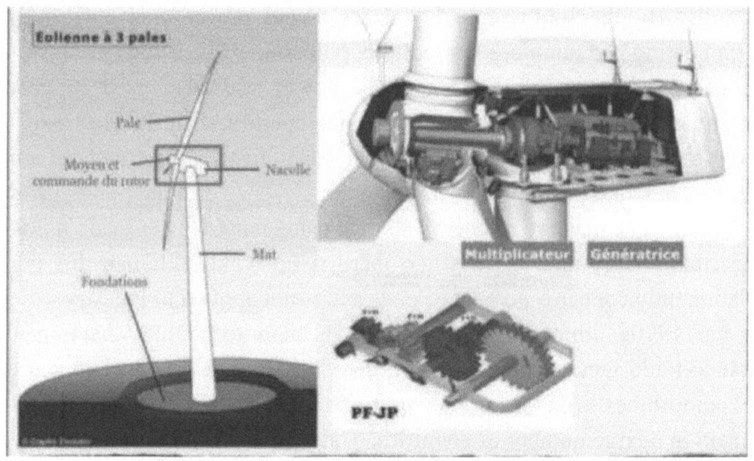

Figure 2.1 : Description of a wind turbine

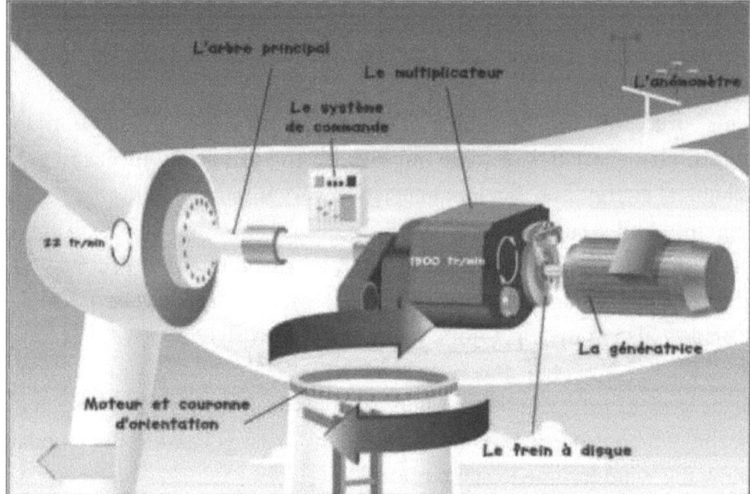

Figure 2.2: Operating principle of a wind turbine

2.3 Characteristics and dimensions

The energy efficiency and power output of wind turbines are a function of wind speed. For three-bladed wind turbines, at the beginning of the operating range (3 to 10 m/s), the power is approximately proportional to the cube of this speed, up to a speed ceiling of 10 to 25 m/s linked to the capacity of the generator. Three-bladed wind turbines operate at wind speeds generally between 11 and 90 km/h (3 to 25 m/s). At

14

beyond that, they are gradually shut down to ensure the safety of the equipment and minimise wear and tear. The wind turbines currently on the market are designed to operate in the 11 to 90 km/h (3 to 25 m/s) range, whether from Enercon, Areva for offshore wind turbines, or Alstom for onshore wind turbines[5] as well as offshore. [6]

Like solar and other renewable energiesenergies renewable , the massive use of wind power requires either a back-up energy source for periods when there is less wind, or means of storing the energy produced (batteries, hydraulic storage or, more recently, hydrogen, methanation orair compressed). [6]

Different types of wind turbine

In all, there are 11 types of wind turbine, 6 of which are presented here:

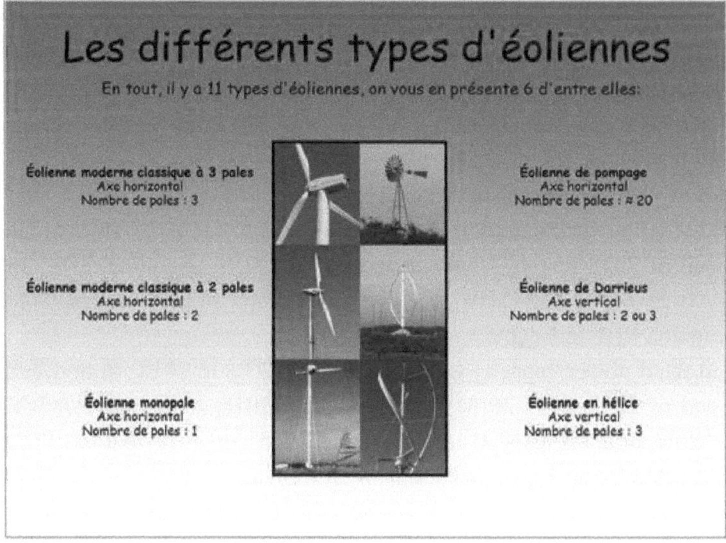

Figure 2.3 : Types **of** wind turbine

The technology currently most widely used to harness wind energy uses a propeller on a horizontal axis. Some prototypes use a vertical axis of rotation: a new technology with a vertical axis is that of the *Kite wind generator* (inspired by kitesurfing) which, to capture the strongest wind possible, uses cables and wings that can reach heights of 800/1,000 m. (ref 6)

Horizontal axis technology has certain disadvantages:

• 1 The spatial dimensions are significant, corresponding to a sphere with a diameter equal to that of the propeller, resting on a cylinder with the same diameter. A high mast is needed to capture the strongest possible wind;

• The wind must be as regular as possible, which means that it is impossible to set up in an urban environment or in very steep terrain;

• the speed at the tip of a blade increases rapidly with its size, with the risk of causing malfunctions and noise to the surrounding area. In practice, the blades of large wind turbines never exceed a speed of around 100 m/s at their tip. In fact, the larger the wind turbine, the slower the rotor rotates (less than 10 revolutions per minute for large offshore wind turbines).

The new wind turbines currently being developed aim to develop a technology that is free of the noise, bulk and fragility of bladed wind turbines, while being able to use the wind whatever its direction and strength. Numerous variants are being studied in real-life trials. Some wind turbines are small (3 to 8 m wide, 1 to 2 m high), with the aim of being able to install them on the flat roofs of residential buildings in towns and cities, or on the roofs of industrial and commercial buildings, in power ranges from a few kilowatts to a few dozen kilowatts of average power. Their rotation speed is low and independent of wind speed. Their power varies with the cube of the wind speed (wind speed increased to power 3): when the wind speed doubles, the power is multiplied by 8. Wind speeds can vary from 5 km/h to over 200 km/h without the need to "feather" the blades. [6]

2.4 Map of wind resources in Africa [6].

Installed wind power capacity in Africa rose by 16.5% in 2019, from 5,728 MW at the end of 2018 to 6,673 MW at the end of 2018, including 2,085 MW in South Africa and 1,452 MW in Egypt. Additions in 2019 were 944 MW, including 262 MW in Egypt.

Installed wind power capacity in Africa rose by 20% in 2018, from 4,758 MW at the end of 2017 to 5,720 MW at the end of 2018, including 2,085 MW in South Africa and 1,190 MW in Egypt. Additions in 2018 totalled 962 MW, including 380 MW in Egypt and 310 MW in Kenya.

It rose by 16% in 2017 (12% in 2016, 30% in 2015, 58% in 2014), from 1,612 MW at the end of 2013 to 2,536 MW at the end of 2014 and 3,488 MW at the end of 2015,

3,917 MW at the end of 2016 and 4,538 MW at the end of 2017; more than half of the leap forward of 934 MW in 2014 took place in South Africa: +560 MW and almost a third in Morocco: +300 MW; in 2015, South Africa accounted for 64% of the increase in the African equivalent with +483 MW, followed by Ethiopia: +153 MW; in 2016, all the commissionings took place in South Africa: +418 MW; similarly in 2017: +621 MW.

South Africa ranks 1[er] with 2,085 MW installed at the end of 2018, or 36% of the African total, compared with 1,473 MW at the end of 2016, 1,053 MW at the end of 2015, 570 MW at the end of 2014 and 10 MW at the end of 2013 [95]; After taking a decade to install its first 10 MW of wind power, in 2013 it was developing 3,000 MW to 5,000 MW of wind power projects, with 636 MW under construction and 562 MW close to financial close; the *Power Sector Integrated Resource* Plan 2010-2030 envisages 9,000 MW of wind power by 2030.

The most advanced project is the Sere wind farm, built by the national electricity company Eskom on the west coast 300 km north of Cape Town; its 106 MW capacity (46 Siemens 2.3 MW turbines) will enable it to produce 240 to 300 GWh/year (load factor: 26 to 32%).

Energy production in Africa [8]

Since 2010, wind power has become the second most important source of renewable energy after solar power.

Table 2.1: Energy production in Africa

Source	1990	%	2000	%	2010	%	2011	2012	2013	2014	% 2014	change 2014/1990
Chartion	100.2	87.5	126,9	87.2	143,9	87.8	142.7	146.0	145,0	147.5	87,6	+47 %
Oil	0		0,9	0.6	0.5	0.3	0.5	0.3	0.2	0.2	0,14	ns
Natural gas	1.5	1.3	1.4	1.0	1,3	0.8	1.1	0,96	1,02	0,87	0,5	-42 %
Total fossil fuels	101,7	888	129,3	88.8	145,7	88.9	144.3	147,2	146,3		88,3	+44 %
Nudeaire	2.2	1.9	3.4	2.3	3.2	1.9	3.5	3.1	3.7	3.6	2,1	+63 %
Hydraulics	0,09	0.08	0.1	0.07	0,18	0,11	0,18	0,10	0,10	0,08	0,05	-3%
Biomass-waste	10.6	9.2	12,9	8,8	14,9	9.1	15.1	15.3	15.6	15.8	9,4	+49 %
Solar, wind, geoth.	0		0		0,07	0.04	0,07	0,08	0,11	0,29	0,17	ns
Total RE	10,7	9.3	13,0	8,9	15,1	9.2	15,3	15,5	15.9	16,2	9,5	+52 %
Total	114,5	too	145.6	100	154.0	100	163,2	165,9	165,72	168,3	100	+47 %

Growth rate: [8]

Growth in the wind power sector is also very strong (+26.1% on average per year).

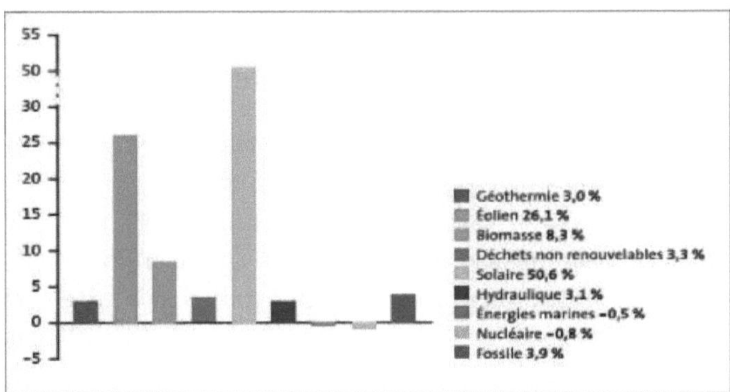

Figure 2.4: Growth rate

Map of Algerian wind farms [8].

Algeria has great ambitions for renewable energy. In addition to the installation of several wind farms in the Hauts Plateaux and the South, there are also wind farm projects in the pipeline:

On 3 July 2014, Algeria inaugurated its first wind farm, located at Kabertene in the centre of the country, north of the town of Adrar. Equipped with twelve wind turbines supplied by the Spanish group, it has a capacity of 10 MW. Total cost: 280 billion.

21 areas offering adequate speed for the siting of wind farms have been identified in anticipation of the siting of future wind farms in 2017.

2.5 Wind farms and power [7]

A **wind farm** is a site where several wind turbines are grouped together to produce electricity. It is generally located in a place where the wind is strong and/or regular. A wind farm on land is made up of several turbines spaced at least 200 m apart, whose electricity output is intended for sale to the local distributor. Although each machine has a small footprint, an area of around 10 hectares is required for a significant wind farm. There are two types of wind farm, onshore and offshore (off the coast).

A 12 MW wind farm, made up of four to six turbines, with a load factor of 23% (i.e. an average production of 24,000 MWh/year), can cover the electricity needs of almost   12,000 people.

(average consumption of 2,000 kWh/year per person), including heating.

The world's largest offshore wind farm is the London Array, comprising **173 turbines** with a capacity of **630 MW**.

The world's largest onshore wind farm is the Roscoe wind farm in the United States (Texas), comprising **627 wind turbines** (781.5 MW) and stretching over 400km^2 . Operator: E.ON Climat and Renewable Energies.

There are two parks:

- Land-based wind farms.
- Offshore wind farms, several kilometres off the coast of Brittany (offshore installations are interesting because they benefit from strong, regular winds).
- Smaller models can also be installed in gardens [8].

The criteria for choosing a wind turbine site depend on the size, power and number of units. They include the presence of a regular wind and various conditions such as the presence of an electricity grid to collect the current, the absence of exclusion zones (including the perimeter of historic monuments, classified sites, etc.), suitable land, etc. [8]

A good wind farm site must have the following qualities:

- sales site
- little turbulence
- easy access
- close to the electricity grid [7]

Power range: [8]

These power ranges are partly determined by the shape and number of blades:
- To produce energy, the wind must have a minimum speed (often 3 m/s, or 10 km/h);
- For safety, if the wind is too strong, the wind turbine is disconnected (often from 90 km/h);

Wind turbines have different shapes and numbers of blades. They therefore have different lift and drag, which explains why they have different optimum operating ranges.

For example: vertical rotors start and stop very quickly, as do wind turbines with many blades (often used as pumping turbines).

2.6 Standards [9]

There are several certifications for wind power in Europe, from Denmark, the Netherlands and Germany. This is due to the fact that their wind energy market is already well established. In France, there is no certification specific to wind power (ISO, AFNOR, etc.). However, there **is standard EN 50 308**: "Aerogenerator, Protection Measures - Requirements for Design, Operation and Maintenance".

This standard has been prescribed by the **European Committee for Electrotechnical Standardization** (CENELEC) on behalf of the European Commission, following the opinion of the "Technical Standards and Rules" Committee, as a "harmonized" standard under the "Machinery" Directive, similar to the **international standard IEC 61400-1**.

It lays down "the requirements for protective measures relating to the health and safety of personnel, applicable to the commissioning, operation and maintenance of horizontal axis wind turbines". wind turbines ".

Its requirements take account of mechanical risks (falls, slipping, etc.), thermal risks (fire, burns, etc.), electrical risks, noise risks and risks arising from failure to observe ergonomic principles. It refers to almost thirty other standards, and in particular to the standards in the EN 292 series (safety of machinery: general principles), which are thus indirectly "harmonised".

2.7 Advantages and disadvantages [9]

The advantages are numerous. Wind energy is clean, renewable, produces no waste and does not pollute. Regular and continuous energy production can be achieved if the wind farm is located in a suitable location with sufficient exposure to the wind.

Wind turbines have a lifespan of around 25 years but, unlike geothermal power stations, they can be dismantled and recycled more easily.

Another **advantage of wind energy** is that the raw material (wind) is free. However, there are also significant disadvantages: the cost of the blades is high, so a long period is needed to ensure a return on the investment.

It should also be noted that the presence of wind turbines on coastal areas (generally those most exposed to the wind) has an unattractive impact on the landscape, and the noise they produce can be annoying. However, these **disadvantages** can be minimised by installing the turbines a few kilometres from the coast, as has been done in Denmark, so as to eliminate the visual and acoustic disturbance.

2.8 Example of a wind turbine installation: Domestic wind turbine [9].

Installing a wind turbine provides you with an income (up to €1,000/year) for 15 years from the electricity produced. However, photovoltaic solar panels are a much more profitable solution. They generate a better income (up to €1,800/year) for 20 years, for an investment 3 times smaller.

Quelle Energie has decided not to offer domestic wind turbines as an energy-saving solution. For the time being, **it is not a sufficiently profitable system**. The cost of the installation is still too high to be amortised by the income generated by the resale of the electricity produced.

A vertical or horizontal wind turbine?

There are currently two types of wind turbine: **horizontal and vertical**. Horizontal wind turbines are the best known: they produce **higher yields** but are also the **most expensive**. Vertical wind turbines are **more affordable**. It is fixed to the roof of your house and has the added advantage of **operating even in light winds**.

Uncertain profitability

Wind turbines are an efficient way of producing electricity cleanly. However, it is not a profitable method of production for private individuals. The **price of a domestic wind turbine is still far too high**. It costs between **€10,000 and €90,000**, including installation. Price variations depend very much on the power used. For example, a 5 kW installation represents an average investment of €30,000, compared with €50,000 for a 10 kW installation. This price also depends on the technology used by the wind turbine. As a result, the income generated by the resale of the electricity produced, which does not exceed €1,000 a year, **does not enable the investment to be amortised**.

More profitable solutions

The low profitability of domestic wind turbines makes them an **unattractive solution**. However, other systems that produce electricity from renewable energy sources offer a better return. This is the case with **photovoltaic panels**, which provide a substantial income (up to €1,800/year) for 20 years. This installation, which **costs 3 times less**, pays for itself in just a few years and guarantees regular income.

To complete your production installation, optimise your electricity consumption with an LED lighting system. Quelle Energie recommends **LED bulbs** because they consume 6 times less energy than conventional bulbs. It is a high-performance lighting solution that is cost-effective, economical, sustainable and ecological.

Installing a domestic wind turbine on the roof of your home

To install a small wind turbine, you don't have to live near an onshore wind farm development zone. All you need is a plot of land exposed to strong, regular and frequent winds. It is then connected to an inverter, which in turn is connected to the grid. Please note, however, that you will need **planning permission and the agreement of your neighbours before** you can install a domestic wind turbine.

The **domestic wind turbine** is a **small wind turbine** optimised to supply electricity to a single dwelling. This type of wind turbine generally has a power output of between 100 W and 20 kW. Installed on a 10 to 35 metre mast,

To install a domestic wind turbine, you don't have to live close to a wind farm development zone. All you need is a plot of land exposed to strong, regular and frequent winds.

Installation takes a few days

Before installation, the craftsman will have to choose the **optimum location** for your future wind turbine and select a size that is as close as possible to your needs. Once this stage has been completed, the professional will be able to install your domestic wind turbine. Installation involves laying a **concrete base** on which the wind turbine is anchored. The wind turbine is then assembled on the ground and lifted using a hydraulic jack. Finally, the installer installs the **inverter** and **connects** it **to the grid**, using a meter to measure the amount of electricity consumed or sold back to the grid.

Conclusion [8]

Wind power is a renewable, non-polluting energy source with great development potential. In fact, wind turbines do not release any gases or dangerous substances into the environment and do not generate any waste.

CHAPTER 3

Hybrid systems

Combining two or more energy sources in the same system brings stability, especially if they are complementary.

A good hybrid system benefits from the combined advantages of its two forms of energy. Flow energies are used to produce most of the energy, at a very low cost, while stock energies are used on demand, as a back-up, to meet exceptional energy needs or to cope with a dip in flow energy production.

Chapter overview: Hybrid systems

3.1 Hybrid Systems

3.2 Tidal turbine

3.3 How a tidal turbine works

3.4 The different types of tidal turbine and their operators

3.1 Hybrid Systems [8]

The recovery of hydraulic energy in its gravitational form has existed for a long time. This is the principle that drives machines and structures such as water mills, tidal mills, hydraulic dams and tidal power plants. On the other hand, the recovery of kinetic energy from river or sea currents was rare before the 21st century[e].

In the early 2000s, the need to develop renewable energies put the spotlight on marine energies, and in particular on tidal power. From the years 2005-2010, the technical maturity of the sector allowed the simultaneous launch of technical and environmental studies around the world.

At the time, tidal turbines were benefiting from huge technical and financial efforts, as had been the case with wind power a few years earlier. At that time, the industrial sector was expected to develop rapidly, particularly for large-scale tidal turbines (machines of 1 MW or more).

The development of new materials (composites, composite concrete, metal alloys, etc.) reinforces the idea that relevant technical solutions adapted to the marine environment could become essential. The 2010s saw the production of demonstrators and prototypes tested all over the world. At the same time, mini-hydro turbines, which are more suited to rivers, easier to maintain and therefore less expensive to invest in, are coming on stream.

The technical difficulties associated with the marine environment, such as corrosion, encrustations and the cost of maintenance and repair operations, contribute to a cost per MWh that is prohibitive compared with other renewable energies. Particularly onerous regulatory constraints are holding back the development of a sector that is already struggling to get off the ground, and the emergence of an industrial sector for maxi-hydro turbines in the near future (2020-2025) is not a foregone conclusion.

In July 2018, Naval Energie announced that it was ending its investment in tidal turbines and would now focus its activities on floating wind turbines and ocean thermal energy. This Naval Group subsidiary had invested €250 million in tidal turbines since 2008 and had just inaugurated the Cherbourg plant dedicated to assembling tidal turbines on 14 June 2018. This decision is justified by the lack of commercial prospects and by a subsidy system that does not provide direct aid to manufacturers during the development phases. The UK's decision not to subsidise tidal turbines, combined with Canada's sensitivity to the cost of the technology, reinforced the view that the market was unprofitable. Placed into receivership by an Irish court, OpenHydro is not expected to fulfil orders for two machines for Japan and Canada.

3.2 Tidal turbine

The operating principle [8]

In practice, the hybrid electric system is centred around the batteries, which are the heart of the system. All the energy sources charge the batteries independently of each other. Some energy generators produce a DC voltage that recharges the batteries directly (DC bus). This is the case with solar panels.

Other appliances produce an alternating voltage, or AC voltage, which must pass through the inverter/charger to recharge the batteries (AC bus). This is the case, for example, with a generator, the mains or a water turbine.

The generator can be started automatically by an internal relay in the inverter/charger if the batteries are too low, for example during periods when there is no sun or wind.

Dimensioning [8]

Sizing a hybrid system is a delicate task, requiring a good knowledge of the products installed, the site's renewable energy potential and the customer's needs (flexibility, safety, economic performance).

For example, the solar regulator, the wind regulator and the battery charger must be set up in the workshop in relation to each other so that they function correctly in the complete system.

A well-designed hybrid system is undeniably the most reliable and economical energy production system for isolated sites, as it takes advantage of all the benefits of the different types of energy. For example, the regularity of production of solar panels goes well with wind turbines, which can produce at night or on cloudy days.3.2 Tidal turbines

In Senegal, unfortunately, both solar and wind power potential are lower during the winter months, so you have to rely on a diesel or petrol generator to make up for the lack of energy at this time of year. For the rest of the year, however, solar and wind power work very well together to keep the batteries charged and the generator switched off.

3.3 How a tidal turbine works [8].

A tidal turbine is a hydraulic turbine (underwater or afloat) that uses the kinetic energy of marine or river currents, in the same way as a wind turbine uses the kinetic energy of the wind.

Figure 3.1 : Tidal turbine (Sabella D03)

The turbine in a tidal turbine transforms the kinetic energy of moving water into mechanical energy, which can then be converted into electrical energy by an alternator. The machines can take a wide range of forms, from large generators of several megawatts submerged at depth in areas with very strong tidal currents to floating micro-generators in small river currents. There seems to be no limit to the inventiveness of designers in this field.

Recoverable power

The recoverable energy is less than the kinetic energy of the water flow upstream of the tidal turbine, since the water must retain a certain residual velocity for a flow to remain. A basic propeller operating model can be used to estimate the ratio of recoverable kinetic power for a section perpendicular to the moving fluid.

This is the Betz limit, equal to $16/27 = 59\%$. This limit can be exceeded if the fluid flow is forced into a vein of variable cross-section (venturi effect) rather than flowing freely around the propeller.

The theoretical maximum recoverable power of a tidal turbine can be expressed as follows:

$$P_{max} = \frac{16}{27} \cdot \frac{1}{2} \cdot \rho. S. V^3 = 295.S. V^3$$

(3.1)

Pmax= power in (W) ;
S= area swept by the blades (m^2);
V= velocity of the water in (m/s)

Tidal turbines take advantage of the density of water, which is 832 times greater than that of air (approximately 1.23 kg m^{-3} at 15°C). Despite a lower fluid velocity, the power that can be recovered per unit of propeller surface area is much greater for a tidal turbine than for a wind turbine.

On the other hand, the power of the current varies with the cube of the speed, so the energy produced by a 4 m/s current is 8 times greater than that produced by a 2 m/s current. Sites with strong currents (> 3 m/s) are therefore particularly favourable, but unfortunately quite rare.

Deformation of the water vein

The incompressibility of water means that the flow through the tidal turbine must be identical upstream and downstream. This means that the product of velocity and cross-section is constant before and after the propeller. As the fluid passes through the turbine, it slows down and the flow widens.

3.4 Different types of tidal turbines and operators [8].

Numerous tidal turbine concepts have been developed, but none has really established itself, each with its own advantages and shortcomings. A few of these have led to demonstrators or experimental projects, but few have reached the stage of industrial production. EMEC lists more than 50 different technical principles, but the European Marine Energy Centre recognises six main types of tidal energy converter. These are horizontal axis turbines, vertical axis turbines, oscillating hydrofoils, venturis, Archimedean screws and kites.

Electricity production from river currents uses mini or micro wind turbines, which are only slightly submerged, have a reduced impact on aquatic fauna and limited investment budgets.

These turbines produce less electricity than conventional turbines, but are much lighter and require much less investment. (Figure 3.2)

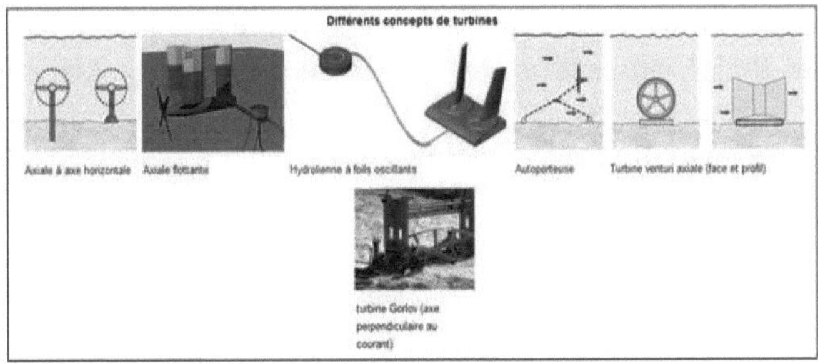

Figure 3.2 : The different types of tidal turbine

Horizontal axis axial turbine

It's the same concept as a wind turbine, but operating under the sea. Numerous prototypes and projects are in operation around the world. The turbines have a single rotor with fixed blades. The power of this type of turbine varies from a few watts to several megawatts. They have between two and a dozen blades, depending on their size. The effects of drag at the tip of the blade limit the speed of rotation of these turbines. Depending on the model, the turbine may or may not be oriented in the direction of the reversing tide. In the case of a fixed turbine drawing its production from reversible current, the profile of the blades is symmetrical to adapt to the two directions of the current

Vertical axis turbine

Misnamed because they can be arranged horizontally or vertically, these turbines have an axis *perpendicular* to the direction of the current. Invented by Georges Darreius in 1923 and patented in 1929. Soviet scientist M. Gorlov perfected it in the 1970s. The blades are made up of 2 to 4 helicoidal foils (curved profile in the shape of an aircraft wing). EcoCinetic, based in La Rochelle, markets a vertical-axis micro-hydro turbine based on the Savonius rotor and designed for use on rivers.

Venturi turbine

The flow of water is guided through a sheath or duct, the cross-section of which narrows at the generator inlet, accelerating the flow and increasing the power available.

Oscillating tidal turbine

Another means of recovering energy that does not require a turbine, bioinspire is based on the movement of membranes or foils oscillating in the current. This type of device is made up of moving planes that actuate and compress a fluid in a hydraulic system. The pressure generated is converted into electricity. Others use undulating membranes.

Self-supporting turbine

Variations, such as the *"tidal kite"*, use the sea current to keep a large kite in "underwater flight", attached to the seabed by a cable. The kite supports a turbine that generates electricity. The first *'Deep Green'* (3 m wide) was tested in Ireland (at Strangford Lough opposite Ulster) by its designer (Minesto, a Swedish spin-off from Saab).

Advantages

Tidal turbines use renewable energy, do not pollute and do not generate waste (at least during their operational phase);

- Due to the high density of water (800 times that of air), tidal turbines are much smaller than wind turbines for the same power output. They have a limited visual impact and, unlike hydraulic dams, do not require complex civil engineering structures;

- Sea currents are predictable, so future electricity production can be accurately estimated.

\-   The potential of marine currents is significant;

The disadvantages

\-   To prevent the development of algae and other fouling organisms on the tidal turbine, antifouling products that are toxic to marine flora and fauna must be used regularly. Carrying out the operation underwater is not an option, both for technical reasons and because the risk to the environment is such that this type of operation has to be carried out by a boat outside a specially equipped careening area. A regular maintenance operation to remove or extract the tidal turbine from the water and redo its careening is therefore essential;

\-   In turbid waters, due to the presence of sand in suspension (the erosion of propeller blades or moving parts by the sand is very high, requiring maintenance operations on the blades. To make replacement easier, some tidal turbines have a structure that emerges from the water (which can be a problem for navigation) or have ballast systems that allow the production units to be lowered or raised;

\-   Tidal turbines create areas of turbulence, which modify sedimentation and currents, with possible effects on flora and fauna just downstream of their position. These aspects are analysed in the impact studies;

\-   Fish, marine mammals or divers could hit the propellers and be injured more or less seriously. However, the propellers can rotate very slowly (depending on the resistance put up by the alternator and therefore the type of tidal turbine).

CHAPTER 4

Photovoltaic solar energy

The differential amplifier is one of the best directly coupled stages. In its basic form, the op-amp typically consists of at least two differential amplifier stages. The differential amplifier is fundamental to the internal operation of an op-amp, since it determines the input characteristics of a typical op-amp.

Chapter overview: Photovoltaic solar energy

4.1 Principle of a photovoltaic installation

4.2 Algeria's solar potential

4.3 Photovoltaic cell technologies

4.4 Photovoltaic modules

4.5 MPPT

4.6 Photovoltaic characteristics and connectors

4.7 The inverter

4.7.1 Role

4.7.2 Principle

4.7.3 Characteristics and performance

4.8 Example of a photovoltaic installation

4.1 Principle of a photovoltaic installation

Photovoltaic solar energy comes from the conversion of sunlight into electricity within semiconductor materials such as silicon or materials coated with a thin metal layer. These photosensitive materials have the ability to release their electrons under the influence of external energy. This is the photovoltaic effect. The energy is provided by photons (components of light) which collide with the electrons and release them, inducing an electric current. This micro-power direct current, calculated in crete watts (Wp), can be transformed into alternating current using an inverter.

The electricity produced is available as direct electricity or stored in batteries (decentralised electrical energy) or as electricity injected into the grid. A solar photovoltaic generator is made up of photovoltaic modules, which are themselves made up of interconnected photovoltaic cells.

The performance of a photovoltaic installation depends on the orientation of the solar panels and the sunshine zones in which you are located.

The future of photovoltaics in industrialised countries lies in its integration on the roofs and facades of solar homes.

Structure of photovoltaic installations [10]

Adaptation between the photovoltaic generator and the electrical load is achieved using static converters, depending on the particular requirements.

Figure 1.4 shows schematically three cases:
•   The most widespread application is that shown in Figure 1.4.a. This involves charging batteries and supplying independent users. A current variator adapts the variable voltage of the panels to the constant voltage imposed by the battery. This type of energy conversion is suitable for locations that are a long way from the grid, as well as for continuous stand-alone use. Power ratings range from a few 100 W to KW.
•   The second application is the use of solar photovoltaics to pump groundwater. Figure 1.4.b shows a DC inverter and a three-phase inverter feeding an asynchronous motor, driving a pump. Power ratings also range from a few 100 W to KW.
Figure 1.4.c shows the recovery of solar energy from the distribution network. A three-phase or single-phase inverter converts DC energy into AC. Even if this application is still too limited because it is not yet economical, it can become interesting and make a certain contribution to the supply of electrical energy. The power of such an application can reach some 100 KW.

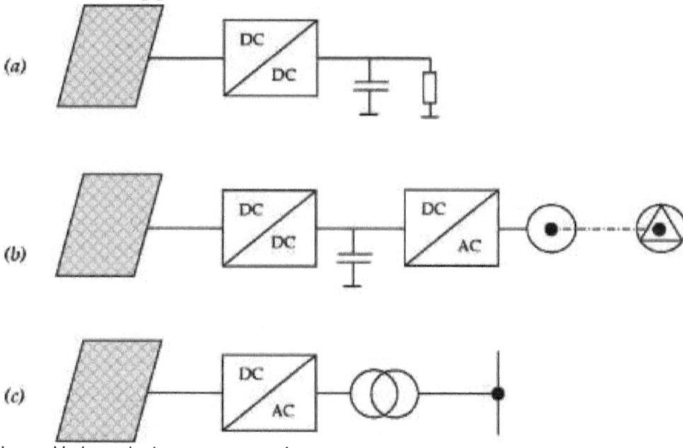

(a) Battery charging and independent use power supply ;
(b) Powering a motor-driven pump unit to pump groundwater ;
(c) Energy recovery in the distribution network.

Figure 4.1: Schematic representation of solar energy conversion
The way in which photovoltaic energy is integrated into electricity systems depends on whether the system is connected to the grid, isolated or hybrid. In each case, storage of the electricity produced by the photovoltaic generator may be necessary for different reasons **4.2 The solar resource in Algeria [14]**
The solar deposit is a set of data describing the evolution of available solar radiation over a given period. It is used to simulate the operation of a solar energy system and to design it as accurately as possible to meet demand.
It is used in fields as diverse as agriculture, meteorology, energy applications

and public safety.

Thanks to its geographical location, Algeria has one of the largest solar deposits in the world.

in the world, as shown in Figure 4.2.

More than 2,000 hours of sunshine over almost the entire country

annually and can reach 3900 hours (high plateaux and Sahara).

The energy received daily on a horizontal surface of 1 m2 is of the order of 5 kWh over most of the country, i.e. almost 1700 kWh/m2/year in the North and 2263 kWh/m2/year in the South.

The country's solar potential exceeds 5 billion GWh.

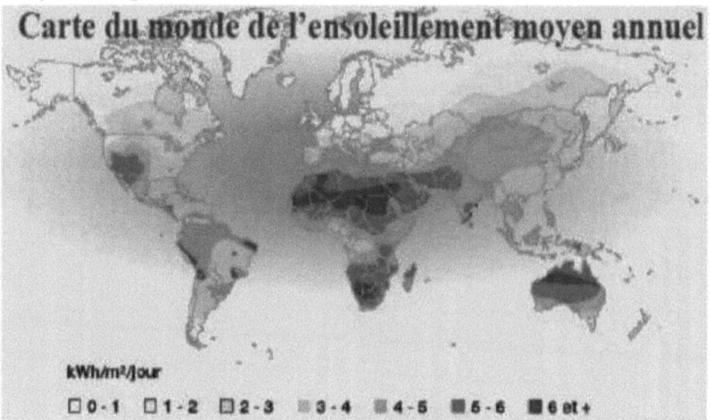

Figure 4.2: World map of annual mean sunshine

Following a satellite evaluation, the German Space Agency (ASA) has concluded that Algeria has the greatest solar potential in the entire Mediterranean basin: 169,000 TWh/year for solar thermal energy and 13.9 TWh/year for solar photovoltaic energy.

Algeria's solar potential is equivalent to the 10 large natural gas deposits that would have been discovered at Hassi R'Mel. The distribution of solar potential by climatic region in Algeria is shown in Table 4.1, according to the amount of sunshine received annually.

Table 4.1: Solar potential in Algeria

Regions	Coastal regions	Hants plateaux	Sahara
Surface area (%)	4	10	86
Average sunshine duration (h/year)	2650	3000	3500
Average energy consumption (kWh/m /year)^2	1700	1900	2650

The average insolation period in southern Algeria is around 3,500 hours/year, the highest in the world at around 9 hours/day (see Figures 4.3 and 4.4), and is still higher than 8 hours/day over most of the country. The southern region, in particular the south-east and south-west, has the greatest potential in the whole of Algeria (see Figures 4.5 and 4.6).

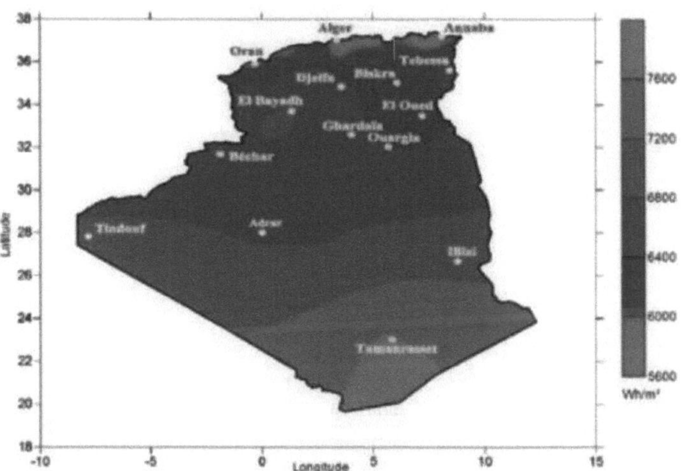

Figure 4.3: Average annual global irradiation received over a horizontal surface Case of a totally clear sky

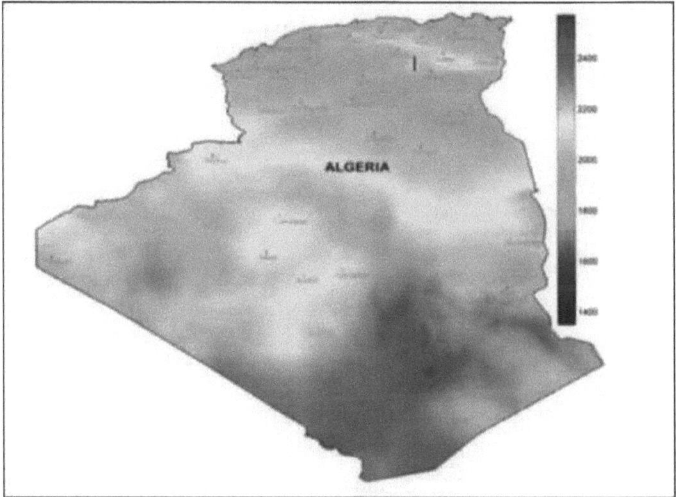

Figure 4.4 : Average annual sums of inclined global irradiation.

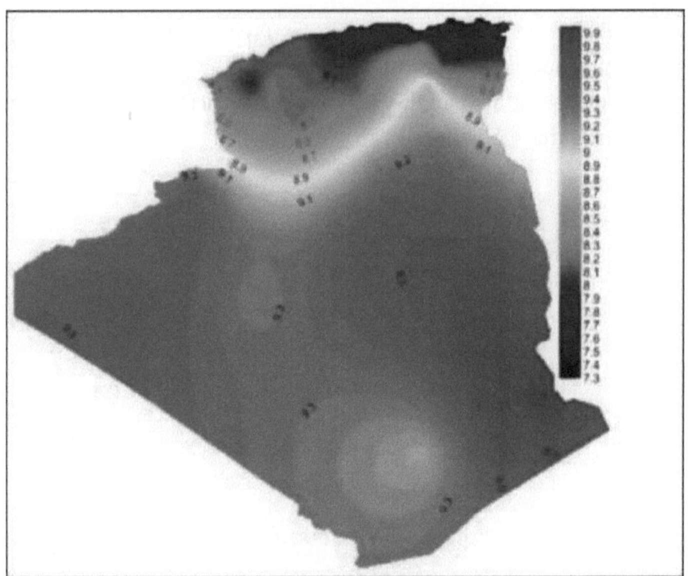

Figure 4.5 : Map of average annual sunshine duration in hours (1983 -2012)

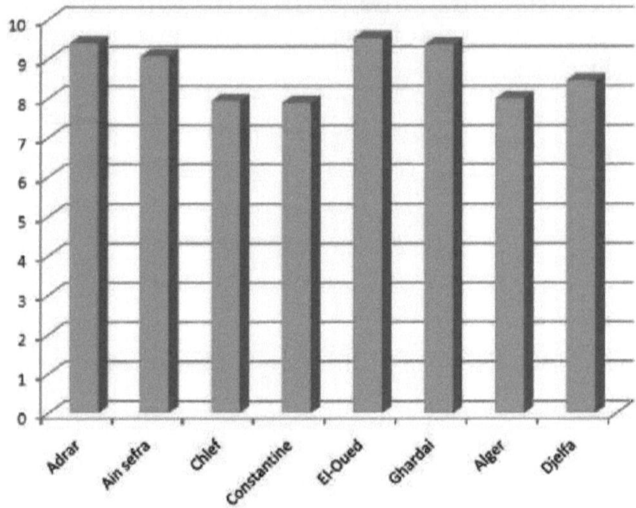

Figure 4.6 : Average annual insolation period measured

4.3 Photovoltaic cell technologies [13]

Photovoltaic solar cells are semiconductors capable of converting light directly into electricity. This conversion, known as the photovoltaic effect, was discovered by E. Becquerel in 1839, but it was not until almost a century later that scientists were able to explore and exploit this physical phenomenon.

Solar cells were first used in space applications in the 1940s. Post-war research led to improvements in performance and size, but it was not until the energy crisis of the 1970s that governments and industry invested in photovoltaic technology and its terrestrial applications.

Today, research laboratories and industry are working together to develop new concepts and processes that can improve the electrical performance and reduce the cost of solar cells. In this way, modern photovoltaic modules, made up of interconnected cells, have amply demonstrated their efficiency and high reliability. What's more, their field of application is constantly expanding, from pumping to lighting, not forgetting all the pocket electronics applications.

The different types of solar cells

There are different types of solar cells (or photovoltaic cells), and each type has its own efficiency and cost. However, whatever their type, their efficiency remains fairly low: from 8 to 23% of the energy they receive.

There are currently three main types of cell:

- Monocrystalline cells: These are the most efficient (12 - 16%; up to 23% in the laboratory), but also the most expensive, because they are so complicated to manufacture.

- Polycrystalline cells: They are easier to design and less expensive to manufacture, but their efficiency is lower: 11% - 13% (18% in the laboratory).

- Amorphous cells: These have a low efficiency (8% - 10%; 13% in the laboratory), but only require very thin layers of silicon and are inexpensive. They are commonly used in small consumer products such as solar calculators and watches.

- The 3eme generation cells currently being developed in the laboratory have a higher output, but have not yet been industrialised.

Our quest for maximum performance has therefore led us to source monocrystalline cells, which offer the best efficiency in real-life conditions.

Performance

Voltage: A photovoltaic cell produces a virtually constant voltage of 0.5V. This corresponds to the cut-off voltage of a diode since, as we shall see later, a solar cell can be likened to a PN junction. To obtain a higher voltage, it will therefore be necessary to connect these cells in series, thus forming a "module". However, we must bear in mind that too high a temperature can reduce the voltage supplied by a cell.

Intensity: The intensity supplied by a cell depends on the surrounding luminosity and the size of the solar panel attached to it. The larger the panel, the greater the intensity supplied. The power supplied by a panel is generally measured in "watt-crete", and this under optimum operating conditions, i.e. in the sun, at midday, in cold weather with clear skies (the maximum intensity of the sun at this time is 1,000 W/m).[2]

4.4 Photovoltaic modules

- 1 the photovoltaic effect :

When a material is exposed to sunlight, the atoms exposed to the radiation are "bombarded" by the photons making up the light; under the action of this bombardment, the electrons in the upper electron layers (called valence layer electrons) tend to be "torn off":

If the electron returns to its initial state, the electron's agitation results in the material heating up. The kinetic energy of the photon is transformed into thermal energy.

In photovoltaic cells, on the other hand, some of the electrons do not return to their initial state. The "disconnected" electrons create a low DC voltage. Some of the kinetic energy of the photons is thus directly converted into electrical energy: this is the photovoltaic effect.

The photovoltaic effect is the direct conversion of energy from solar radiation into electrical energy. The term "photovoltaic" comes from the Greek "phos", meaning light, and "voltaic", a word derived from the Italian physicist Alessandro VOLTA, known for his work on electricity.

- The photovoltaic cell : [11] [12]

Principle of operation of a photovoltaic cell :

Let's consider a p-n junction and close its electrodes with a short-circuit. We know that, under these conditions, on either side of the junction, over a small but finite thickness, there is a deserted zone in which there is an intense electric field produced by a potential difference (pd) of a few tenths of a volt. No current flows in the external circuit.

If an incident photon generates an electron-hole pair in the deserted zone, under the effect of the electric field these carriers will move in opposite directions to each other and join the majority carriers of the same sign in the P and N regions. This corresponds to a current flowing through the junction in the difficult conduction direction. If P photons are absorbed per unit of time, and taking into account the inevitable pairs produced in the vicinity of the deserted zone which are likely to diffuse and recombine, the current density will be of the form

$$(4.1)$$

With

K: Boltzmann's constant: $1.38*10^{-23}$ J/°K. e: electron charge: $1.6*10^{-19}$ C.

If an e.m.f. is added to the circuit, the total current drawn will be of the form

$$(4.2)$$

With

J_0 the initial current density.

T: junction temperature in degrees Kelvin

By varying V in magnitude and sign, we obtain the characteristic below. It can easily be plotted by dragging the obscurity characteristic by a quantity.

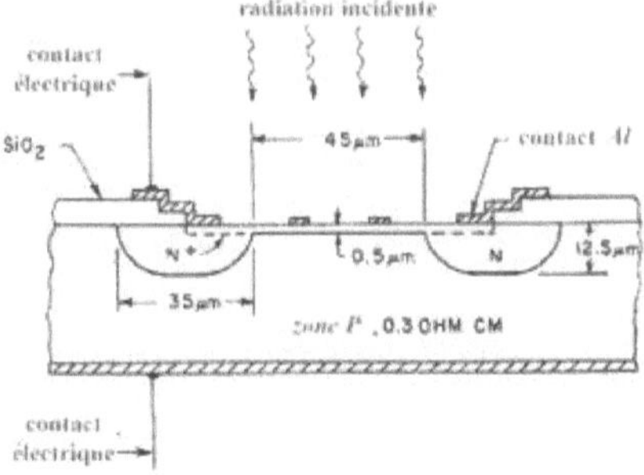

Figure 4.7: Cross-section of solar cell

This effect therefore requires a potential barrier to manifest itself. In practice, a p-n junction is produced, the essential characteristic of which is to be very asymmetrically shaped and above all to have a very thin n zone, so that the space charge zone is very close to the surface, in order to obtain maximum efficiency.

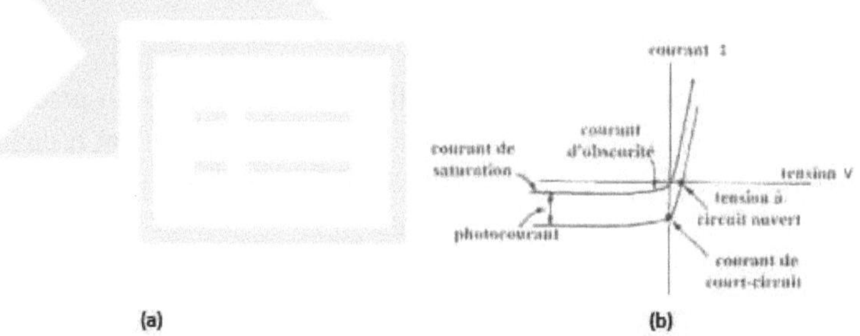

Figure 4.8: a) Principle of the photodiode, b) I=f(V) characteristic of a photodiode

The relationship giving the photocurrent in an open circuit is

$$I = h\, e\, N_1\, G \tag{4.3}$$

Or

h : is the *quantum efficiency,* i.e. the number of excess carriers produced per absorbed photon,

N_1: the number of photons of wavelength 1 absorbed per unit of time G: the photoconductive gain representing the ratio between the lifetime of a carrier and the transit time along the length of the sample, but here with a unit gain since all the carriers generated will cross the barrier.

Furthermore, given the solar spectrum, which has a certain width and a high degree of inhomogeneity, as shown in the figure at the beginning of the chapter, this relationship will have to be integrated over the entire wavelength range.

As in a p-n junction, the direct current is expressed according to the relation

$$I_d = I_s\left[exp\left(\frac{eV}{KT}\right) - 1\right] \qquad (4.4)$$

These cells can be used in two different ways: either in series with a female as an IR radiation detector (Ge photodiodes), or as a generator by closing them off to an R load.

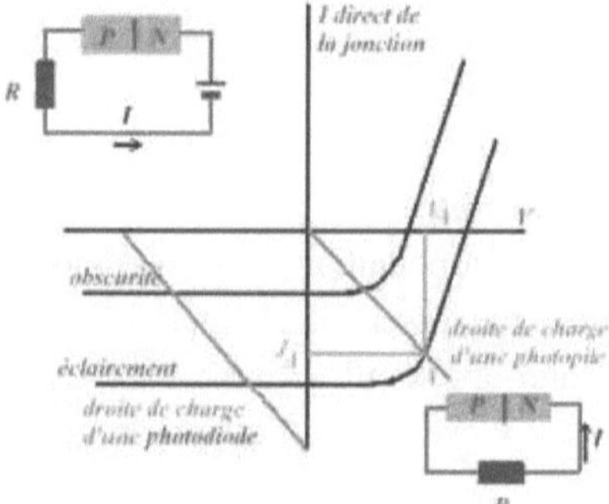

Figure 4.9: The operating zones of a PV cell

In the latter case, sunlight is always used as the excitation energy. In practice, they should be switched on at the optimum load corresponding to point A on the graph, i.e. the resistance R should be such that R = V$_A$/J$_A$, which is unfortunately not often the case.

In all cases, to obtain the best yields, it is necessary for the hole electron pairs to be released in the desertified zone or in its immediate vicinity. This condition can never be perfectly met. The optimum is approached by folding the junction at a minimum distance from the illuminated zone. In addition, the cell resistance (internal resistance of the equivalent generator) must remain low, which limits the extent of the illuminated areas.

The photovoltaic (PV) cell is a non-linear power source that can be represented by the equivalent electrical diagram in Figure (4.5).

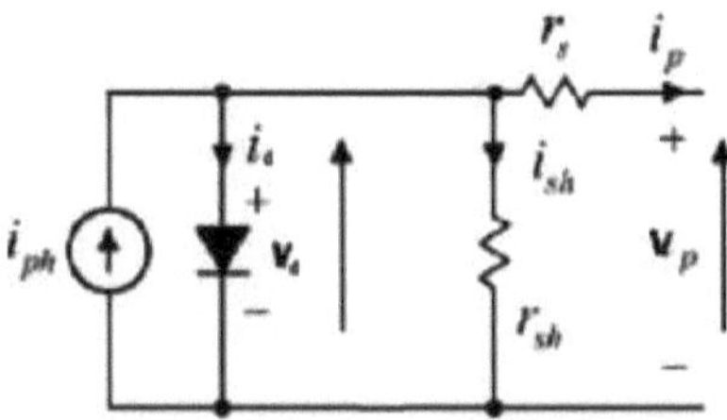

Figure 4.10: Equivalent electrical diagram of the PV cell

With:

rsh : Shunt resistor which takes into account the inevitable leakage of current between the collector grid and the back current of the cells.

rs: Series resistance due essentially to the contact resistance of the collector grids with the cell surface and the resistance of the material making up the cells.

By studying the physics of a solar cell, we can obtain the current equation of the load:

$$i_p = i_{ph} - i_d - i_{sh} \qquad (4.5)$$

With :

ip : The current delivered by the solar cell.

iph: Photo current, created by solar photons of energy greater than the gap.

id: Diode current.

ish: The current flowing through the shunt resistor.

n : Corresponding to the ideality factor, it represents the deviation of the characteristics compared with an ideal diode ;

Or the output current can be expressed by equation (4.7).

$$I_p = I_{ph} - I_0 \left(\exp \frac{q(Vp + R_s I)}{n k T} - 1 \right) - \frac{Vp + R_s I}{R_{sh}} \qquad (4.6)$$

Where Vp is the output voltage.

- The influence of illumination and temperature on the operation of a PV cell [11].

1. The influence of solar irradiance on the operation of a PV cell

The electrical energy produced by a photovoltaic cell depends on the illumination it receives on its surface. The following figure shows the current-voltage characteristic of a solar PV cell as a function of irradiance, at constant temperature and ambient air speed. It can be seen that the voltage Vmax, corresponding to the maximum power, varies very little as a function of irradiance, unlike the current Imax, which increases sharply with irradiance.

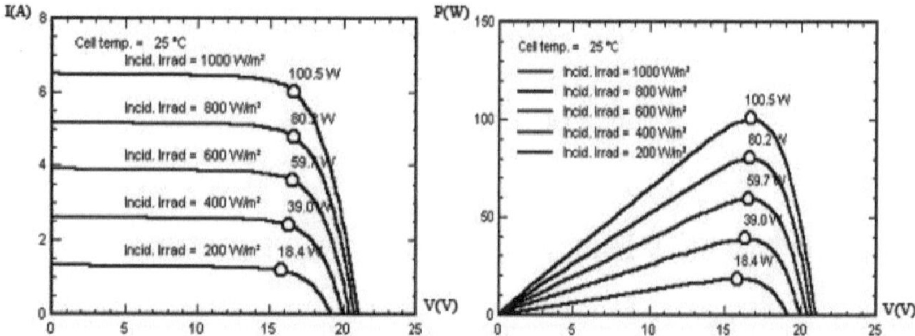

Figure 4. 11: The influence of solar irradiance on the operation of a PV module

2. The influence of temperature on the operation of a PV cell

The electrical characteristics of a PV cell depend on the junction temperature at the exposed surface. The behaviour of the PV cell as a function of temperature is complex.

In the case of silicon cells, the current increases by around 0.025 mA / cm^2 . °C while the voltage decreases by 2.2 mV / °C. The overall drop in power is around 0.4% / °C. So the higher the temperature, the lower the performance of the cell.

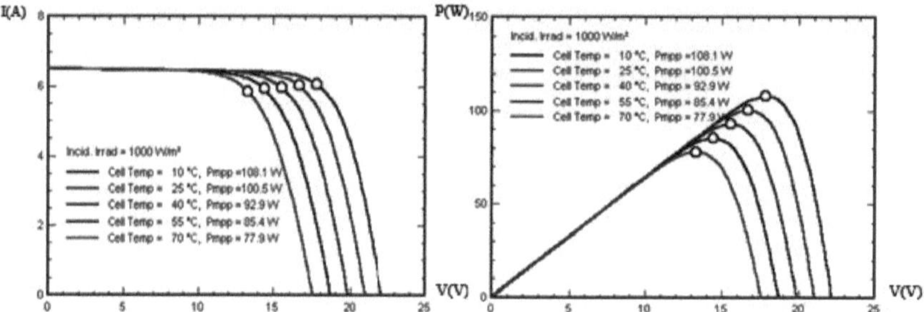

Figure 4. 12: The influence of temperature on the *operation of* a PV module
- PV cell associations [11]

In fact, the association of PV cells is analogous to the association of current generators:

> In series, their tensions add up,

> At the same time, their currents add up.

However, their functioning is altered if one of the associated cells is obscured (shadow, for example).

1. **Serial association of PV cells**

2. If cells are assembled in series, the voltage across the assembly is equal to the sum of the voltages delivered by each of the cells.

$$U = S_{?=1} \, y_c \quad (4.7)$$

U (V), Volt: Terminal voltage of the assembly.

uc (V), Volt: Voltage at the terminals of the cell with index "c".

In this case, the current flowing through the cells is the same, but the cells can operate at different voltages.

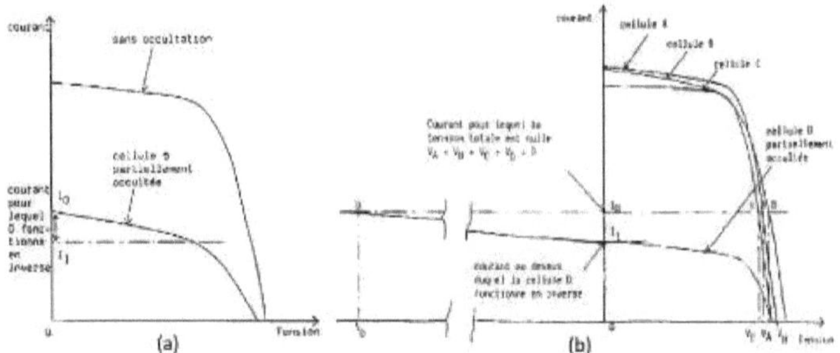

Figure 4.13: a). Operation of cells in series, I-V characteristics of the assembly. **b).** Operation of 4 cells in series, one of them partially occluded.

If a cell is blocked (if it only receives a small part of the solar energy received by neighbouring cells), it can only deliver a limited current. It therefore works in reverse (like a receiver subjected to a voltage that is the opposite of the direct voltage) to the other cells in the module, which deliver a current that exceeds this limit. When operating in this way, the cell heats up and can break down.

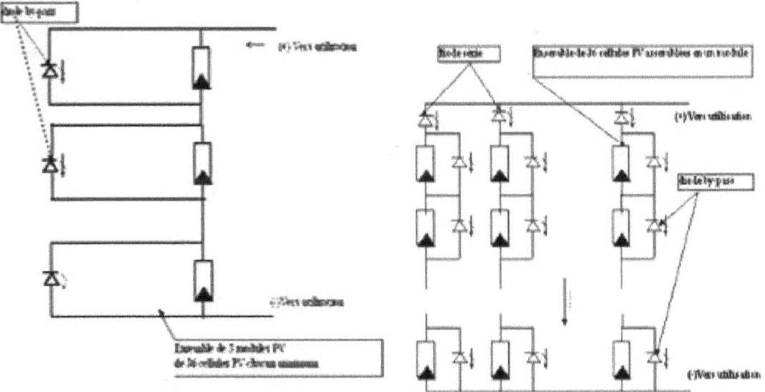

Figure 4.14: Equivalent electrical diagram of the series assembly

Research on this subject has shown that in the case of silicon PV cells, above a reverse voltage of 20 V, the probability of a cell breakdown (destruction of the electrical junction) becomes significant. To limit the maximum reverse voltage that can develop at the terminals of a cell, **manufacturers of photovoltaic modules place a parallel diode, called a bypass diode, every 18 to 36 cells (depending on the application).**

3. **Parallel association of PV cells**

In the case of a parallel circuit, the overall current will be equal to the sum of the currents produced by each cell.

$$I = \sum i = 1/c$$

I(A), Ampere: Current flowing through the assembly.

i_c (A), Ampere: Current flowing in each cell of index "c".

In the case of a parallel combination, the cells deliver the same voltage but can operate with different currents.

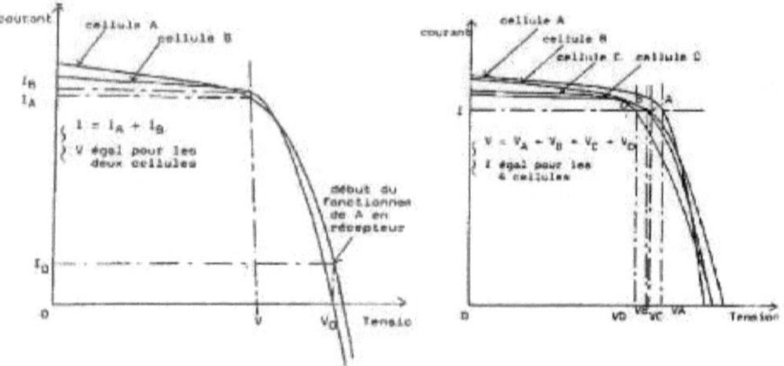

Figure 4.15 : Characteristic curve of a parallel assembly

However, if one or more cells are blacked out, the others become receptive because the operating voltage is higher than the open circuit voltage. Although one cell can dissipate a large current, it is preferable to have an anti-return diode, which also prevents part of the power produced by the cells operating normally from being wasted in another blacked-out cell. To limit these losses and protect the cells, a diode is placed in series, called a series diode, every n cells (n being a function of the characteristics of the circuit).

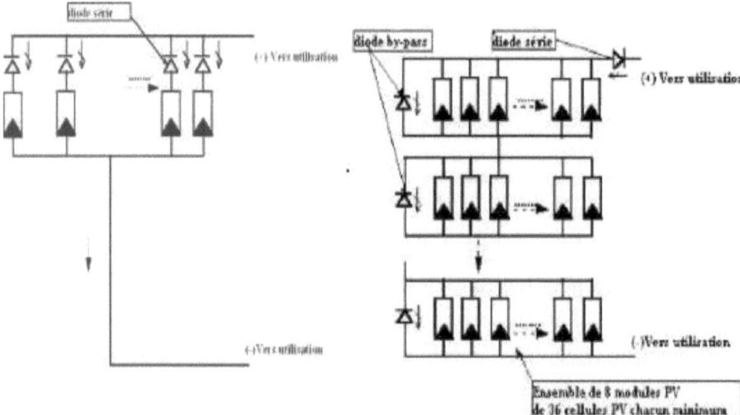

Figure 4. 16: Equivalent electrical diagram of the parallel assembly

In order to limit power losses in these diodes, manufacturers generally use Schottky diodes (e.g. series 1 N 5817, 18 and 19 for currents of the order of 1 A, and SD 51 or MBR 340 M for currents of 20 A or more) whose voltage drops in direct operation are often less than 0.4 V (11).

- Use of bypass and series diodes [11].

The by-pass diodes for protecting cell combinations in series are fitted :

> By the manufacturers when cells are assembled in series (to protect the cells). They will be inserted into the cell connection boxes on each module when they are manufactured.

> By installers when combining modules in series in junction boxes. Parallel association protection diodes:

> Not fitted by manufacturers when cells are combined in parallel (it is not necessary to protect a hidden cell in a module given the low currents involved).

> Are fitted by installers when combining modules in parallel in junction boxes for concealed modules in a set of modules.

4.5 MPPT (Maximum Power Point Tracking)

There are around twenty methods for tracking the maximum power point of a field of modules (Maximum Power Point Tracking), with varying degrees of efficiency and speed.

The power converter control unit ensures that the PV generator operates at the optimum operating point (maximum power point or MPP*) to guarantee maximum electrical power production.

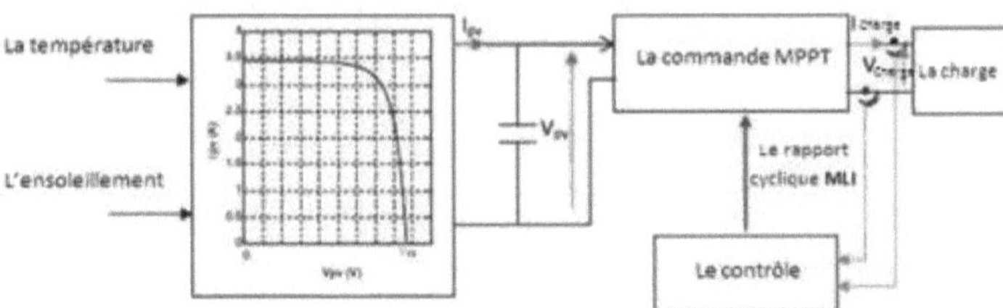

Figure 4. 17 : MPPT control of output parameters

The two most commonly used methods are HillClimbing and P&O (Perturb and Observe). These two methods work on the same principle, which consists of disturbing the operation of the system and then analysing how the system reacts to this disturbance: modifying the chopping duty cycle for the Hill-Climbing method, modifying the voltage at the terminals of the photovoltaic module array for the P&O method. Modifying the conversion efficiency of the inverter disturbs the direct current coming from the modules and consequently the voltage at their terminals and the instantaneous power delivered.

These two methods are therefore based on the control of the instantaneous power delivered by the PV module array as a function of variations in the DC voltage across the PV array (flowchart1 and figures 4.18).

Flowchart 1: Principle of the Hill-Climbing and
P&O algorithm

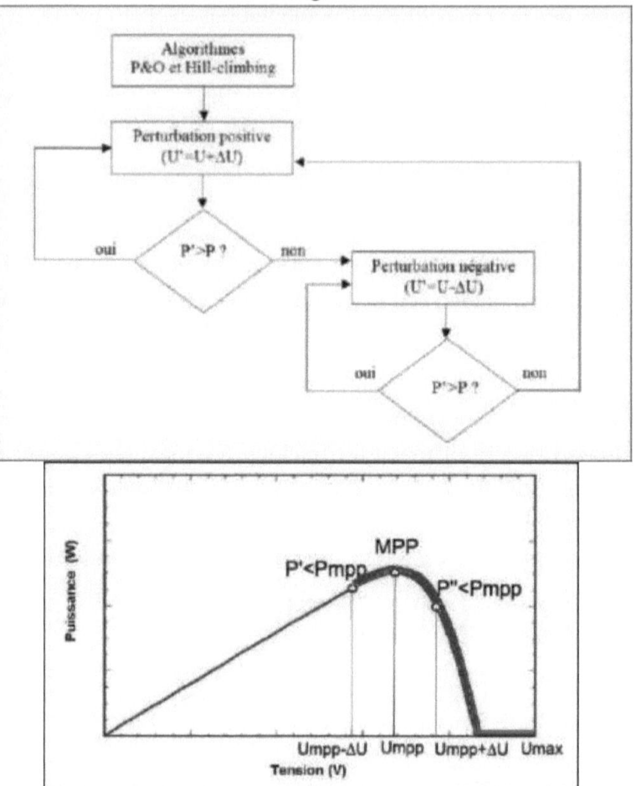

Figure 4.18: Illustration of the Hill-Climbing and P&O algorithm

Other methods for finding the Maximum Power Point (MPP) include :

- The IncCond (Incremental Conductance) method, which consists of comparing the incremental conductance AI/AU with the ratio -I/U: AI/AU= I/U at the MPP, >-I/U to the left of the MPP and <-I/U to the right of the MPP.
- The fractional open circuit voltage method, based on the proportionality between the open circuit voltage (Vco) and the voltage at the point of maximum power (Vmpp).
- The fractional short-circuit current method is based on the proportionality between the short-circuit current (Icc) and the current at the point of maximum power (Impp).
- Fuzzy Logic Control of MPP
- MPP control based on neural networks
- The RCC (Ripple Correlation Control) method, which consists of analysing

the interference caused by the inverter on the voltage and current of the PV module array.

- The Current Sweep method, which periodically calculates the voltage-current characteristics of the PV module array and determines the MPP.

- The DC Link Capacitor Droop Control method

- The Load Current or Load Voltage Maximization method is based on the fact that if the inverter's output power is maximum, then the PV module array is operating at its MPP*.

- The dP/dV or dP/dI Feedback Control method, which consists of analysing the slope of the power-voltage curve of the PV module array (zero at the point of maximum power, positive before the MPP and negative afterwards).

4.6 Photovoltaic characteristics and connectors [11]

Under illumination, the I(V) characteristic of the photodiode no longer passes through the origin of the coordinates. There is a region in which the product (V*I) is negative, and the diode supplies energy. If we restrict ourselves to this active region and count the reverse current positively,

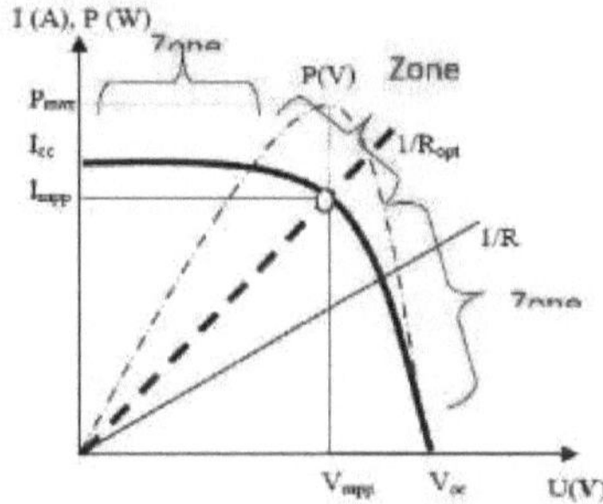

Figure 4.19: The I(V) characteristic of a photovoltaic module

The *I(V)* characteristic of a photovoltaic generator can be broken down into 3 zones Figure (4.) :

• A zone similar to a current generator *Icc* proportional to irradiation. The internal admittance can be modelled by: (1 / Rsh) (*Zone 1*).

• A zone similar to a voltage generator *vco* with an internal impedance equivalent to *Rs* (*Zone 2*).

• A zone where the internal impedance of the generator varies very sharply from *Rs* to *RSH* (*Zone 3*).

Zone 3 is the operating point at which the power supplied by the generator is at its maximum. This point is referred to as the optimum power point, characterised by the torque (*Impp, Vmpp*), and only a load whose characteristics pass through this point can extract the maximum power available under the conditions in question, hence the concept of TRACKING THE OPTIMAL POWER OF THE SOLAR PANEL.

4.7 The inverter [13]

For more than 10 years now, the global market for photovoltaic systems has been growing at a very high rate of around 30 to 40% a year. This exceptional growth, which is mainly due to photovoltaic systems connected to the electricity distribution grid, is of course reflected in technological innovations and lower costs for photovoltaic modules, as well as major research and development efforts in the field of power electronics.

The technical performance and reliability of the inverters used to connect the photovoltaic modules to the electricity grid can have a major impact on annual electricity production and therefore on the financial profitability of a system.

4.7.1 Role

An inverter is a device for transforming DC electrical energy into AC. They are used in electrical engineering to :

- Either supply alternating voltages or currents of variable frequency and amplitude.

E.g.: This is the case with inverters used to supply AC motors that have to run at variable speed, for example (the speed is linked to the frequency of the currents flowing through the machine).

- Either supply one or more AC voltages of fixed frequency and amplitude.

E.g.: This is particularly the case for emergency power supplies designed to replace the mains in the event of a mains failure, for example. The energy stored in the back-up batteries is restored in DC form, so the inverter is needed to recreate the mains voltage and frequency.

A distinction is made between voltage inverters and current inverters, depending on the DC input source: voltage source or current source. Voltage inverter technology is the best known and is used in most industrial systems, in all power ranges (from a few Watts to several MW).

4.7.2 Principle

Inverters designed for photovoltaic systems are somewhat different from the conventional inverters used in electrical engineering, but the AC/DC conversion objective is the same. The main feature of the PV inverter is the search for the system's best operating point.

In fact, the PV generator (set of PV modules) has a non-linear IV characteristic curve (Figure 4.20).

For a given illumination and temperature, the voltage in an open circuit or with a high load is more or less constant (similar to a voltage source), whereas in a short-circuit or with a low load the current is practically constant (current source). The generator is then neither really a source of voltage nor really a source of current either.

The open-circuit voltage is sensitive to temperature and decreases as the temperature rises. The short-circuit current is proportional to the illumination: it increases as the illumination increases.

The system's best operating point corresponds to the point on this curve where the power, the product of voltage and current, is maximised. It is located in the middle of the characteristic (Figure 4.20).

Under steady state conditions, the transducer voltage and current are assumed to be constant. The use of a voltage inverter rather than a current inverter is then essentially motivated by technological reasons.

The voltage inverter imposes a system of voltages at its output in the form of pulse-width modulated (PWM) peaks. These crests pose no problem for powering a motor, but are incompatible with sinusoidal mains voltages.

An inductance is then placed between each inverter output and each mains phase (single-phase or three-phase inverter), which acts as a filter and enables the inverter to supply quasi-sinusoidal currents to the mains: from a formal point of view, it transforms the voltage inverter into a current inverter (Figure 4.21).

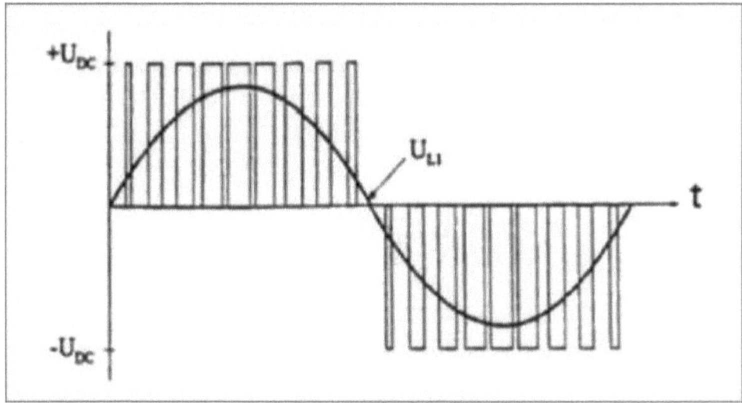

Figure 4.20: Voltage filtering by the output inductor

UDC corresponds to the voltage across the input capacitor of a simple circuit (figure 4.20) and UL1 to the voltage injected into the mains, i.e. 50 Hz frequency.

Main types of inverters available

Inverters are bridge structures usually consisting of electronic switches such as IGBTs (power transistors). In the standard case, by means of a set of appropriately controlled switches, most often using PWM, the DC electrical energy supplied is modulated to obtain an AC signal at the mains frequency.

There are many electronic circuits for converting electrical energy:

The simplest circuit consists of thyristors. This technology was used in the first PV inverters (and is still available in single-phase and three-phase).

Although inexpensive, it has a more or less rectangular output current that induces reactive power and harmonics that affect the inverter's efficiency and can disturb the grid.

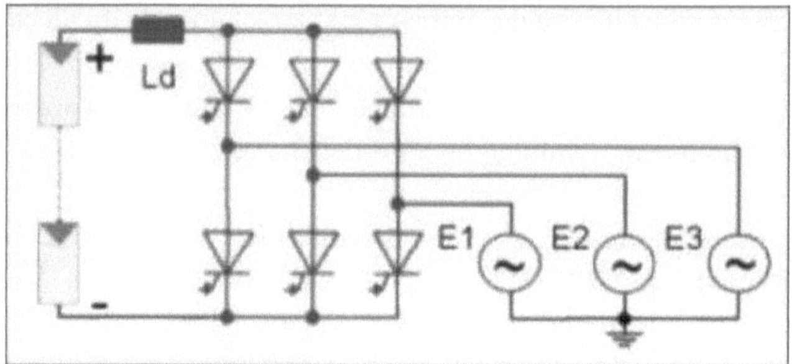

Figure 4. 21: Circuit using thyristors

Figure 4.22 shows an example of a simple circuit consisting of a PWM-controlled transistor bridge. The alternating signal obtained is then filtered by the inductance Ld located before the transformer (or Lac on the other diagrams) in order to obtain a sinusoidal alternating signal at the mains frequency.

This signal is then adjusted to the mains voltage by a 50 Hz transformer, which also provides galvanic isolation for the circuit.

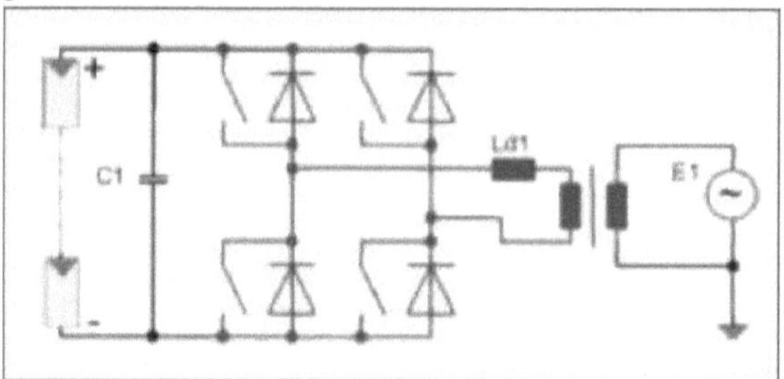

Figure 4.22: Simple circuit using a transistor bridge

To work with a wider range of input voltages, a boost converter can be added to the input of the bridge.

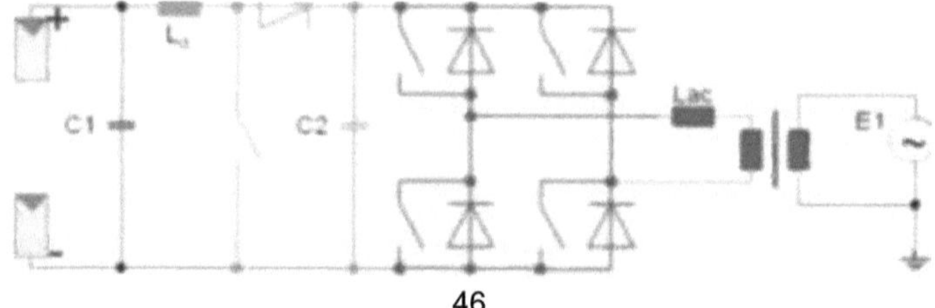

Figure 4.23: Transistor bridge circuit with boost converter

The assembly shown in figure 4.24 comprises 3 different stages. It consists of a high-frequency transformer which adapts the input voltage while reducing the weight of the inverter. The output signal is alternating. A rectifier then converts it to DC. The output bridge uses amplitude modulation to transform the DC signal into a sinusoidal AC signal adapted to the mains frequency.

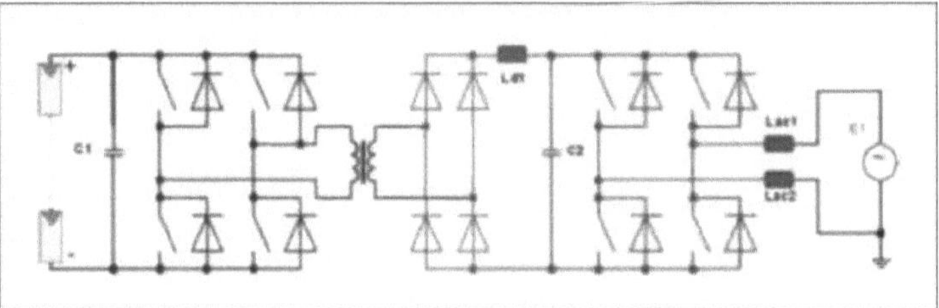

Figure 4.24: 3-stage circuit with high-frequency transformer*.

The circuit shown in figure 4.25 consists of 4 stages. This circuit requires the control of 7 switches, compared with 8 for the circuit in figure 6. It consists of a step-down converter, a push-pull circuit followed by a rectifier, and an output bridge.

The "step-down converter + push-pull transformer" allows the input voltage to be adapted. This allows the inverter to have a wider range of possible input voltages, and therefore greater flexibility when combining with PV modules. The rectifier 'rectifies' the voltage at the push-pull output, and the output bridge uses amplitude modulation to transform this DC signal into a sinusoidal AC signal adapted to the mains frequency.

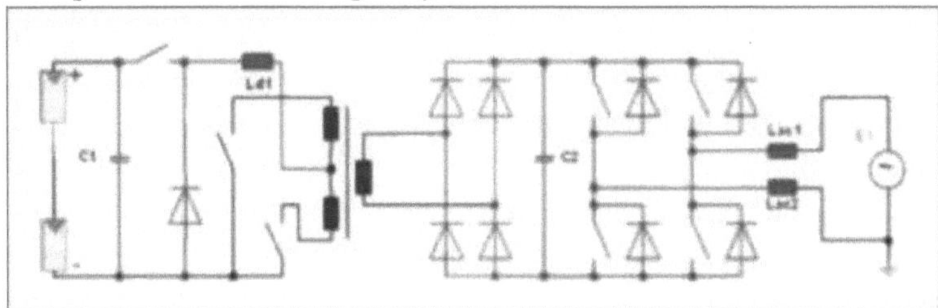

Figure 4.25 : 4-stage push-pull circuit

Finally, the circuit shown in Figure 4.26 is a simple example of transformerless inverter technology. By eliminating the transformer, which generates significant losses in the circuit during power conversion, efficiency can be increased.
However, we must take into account the problems of electromagnetic compatibility that the transformer eliminated by galvanic isolation.

In this circuit, S1 (for positive and negative currents) and S2 (for positive currents) are controlled at high frequency and the other switches at 50Hz (mains frequency). For higher input voltages, S1 can be controlled alone at high frequency and the other 4 at 50Hz to form a down converter and a push-pull converter.

In both cases, the disadvantage of this circuit is the very high voltage applied across the switches.

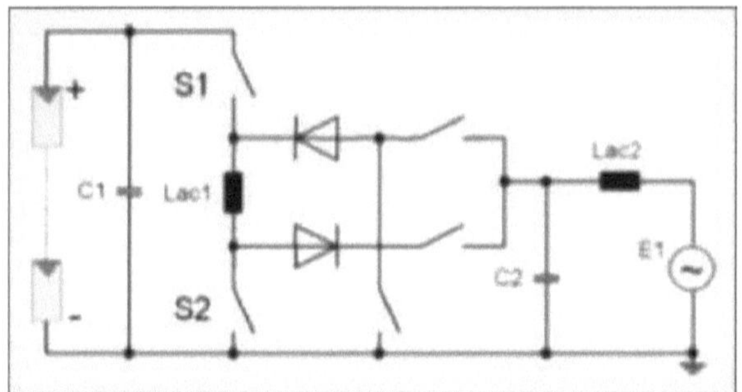

Figure 4.26: Transformerless* inverter circuit with Karschny topology

Role of the input capacitor

All the circuits presented have a high-capacity C1 reference capacitor at the input, which is very important for photovoltaic inverters in maintaining a stable operating point.

It acts as an energy store and filters out voltage fluctuations caused by switching. In this way, it ensures that the current flows uniformly from the PV generator to the grid, keeping the voltage constant.

Aluminium electrolytic (electrochemical) capacitors are used in direct current or very low frequency applications and are those most commonly found in PV systems. They have high capacitance values and low series resistance.

4.7.3 Characteristics and performance

Maximum output :

The efficiency of inverters has been increasing steadily in recent years. This improvement is, of course, helping to drive down the cost of PV-generated electricity.

15 years ago, 90% was considered to be very good efficiency for PV systems. Today, the best inverters reach peak efficiencies of 98% and the average is 95.2%.

The other striking improvement is the "European Efficiency", which takes into account the inverter's part-load efficiency. Because of poor part-load efficiency, the European Efficiency value is lower than the maximum efficiency value. 15

years ago, it could be as much as 5% less than the maximum efficiency, whereas today the difference is between 1 and 2%, for the best models.

The average European yield for inverters on the market in 2007 was 94.4%.

From a technical point of view, it should be possible to reduce this gap to 0.5% by optimising part-load efficiency.

Maximum efficiency is also set to rise to 99% over the next few years. An increase of 1% (from 98% to 99%) in efficiency means heat losses halved by 2, which is extremely important for improving the lifespan of components and therefore inverters.

Less heat loss also means that cooling systems are no longer required and inverter casings can be reduced in size.

This improvement in efficiency can be achieved by optimising the components used to minimise heat loss.

Research into the first nanotube power transistors is continuing. They could replace IGBT transistors and considerably reduce losses, which are already low with IGBTs.

One way of reducing losses in the transistors is to mount several of them in parallel rather than just one, in order to reduce losses during part-load operation.

Input voltage range

Some manufacturers have chosen to extend the range of inverter input voltages as a means of improving their new products. A wide input range makes it easier to choose the inverter when sizing the system, and facilitates stock management for the manufacturer.

One PV module more or less in the system no longer necessarily calls into question the choice of inverter.

However, it is not sufficient to have a generator output voltage within the inverter input voltage range to achieve maximum efficiency. It is possible to optimise the relationship, as shown in the graphs in figure 12.

However, it is not possible to state that a higher voltage is more suitable than a lower voltage, as the response depends too much on the topology of the inverter circuit, as can be seen in figure 4.27.

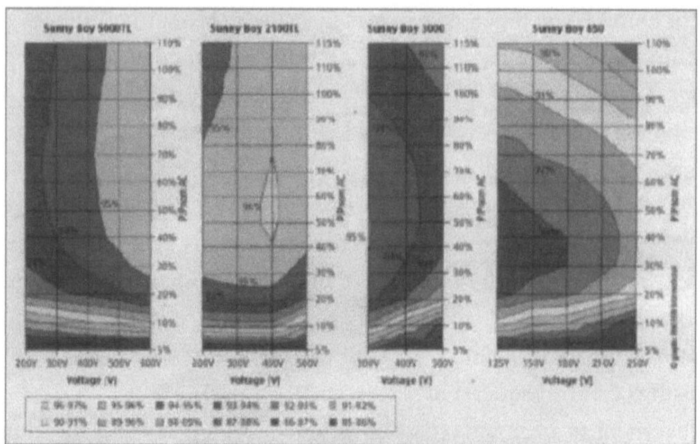

Figure 4.27: Influence of PV generator voltage on inverter input voltage range

Some manufacturers have chosen to use boost converters to extend this input range. These convert small input voltages and high currents into high voltages and low currents. The inverter does not require a powerful transformer. It is easier to improve efficiency when working with low currents because the losses are lower.

are less important, even if the converter itself causes some losses.

However, the manufacturers' enthusiasm is held back by efficiency. In fact, it appears that PV inverters tend to operate with high efficiency in a rather narrow range of input voltages.

So the development of an opposite trend would be quite likely: the development of inverters optimised to work in a reduced range of input voltages but perfectly adapted to the system.

Life expectancy

A 5-year guarantee on products has become the norm with manufacturers, whereas a few years ago it was just 2 years.

It is possible to extend the guarantee to 10 years, or even 20 years depending on the manufacturer. This extended life is the result of using higher quality, oversized components or components that are more resistant to temperature rises.

4.8 Example of a photovoltaic installation [15].

An off-grid photovoltaic installation is one that is not connected to the electricity grid. The energy produced is stored (in batteries) and can then be used to power any type of 12V, 24V or 220V electrical appliance.

Figure 4.28: A photovoltaic installation on an isolated site

Photovoltaic panels generate electricity

Photovoltaic modules convert solar energy (light) into **direct current** electrical energy.

Each solar module has a nominal voltage of 12 or 24 volts, and they are wired in parallel or in series to obtain the required voltage and current. The current is then output at **12, 24 or 48 vol**ts.

The big advantage of this type of energy production is that the transformation of light into electricity takes place without noise or pollution.

Solar regulator automatically regulates battery charge

The solar regulator **regulates the charge of the battery pack to avoid overcharging or discharging** the batteries **too deeply**: - it limits, or even stops, the charging of the solar battery by the photovoltaic module when the battery is fully charged   ;

- it slows down discharging by reducing the load, or even stopping it altogether, to avoid deep discharges that could damage the batteries.

The solar regulator continuously displays the operating status of the photovoltaic panel and the state of charge of the solar battery. This information is displayed either via LEDs, or via an integrated or remote digital display.

Electricity storage in solar batteries

Solar batteries, also known as slow-discharge batteries, store the energy produced by photovoltaic modules so that consumers can be supplied in all circumstances (day or night, clear or overcast sky).

These batteries are specifically designed for solar or wind power applications. They do not have the same characteristics as a car battery, for example.

The capacity of a battery is expressed in Ah (Ampere/hour) and its voltage is 12 Volts (most often), 6V or sometimes 2V (for the largest photovoltaic installations).

The voltage converter or inverter

A converter, or inverter, transforms the direct **current** from the batteries

(12V or 24V or 48V) into 230 Volt alternating current.

To do this, **simply connect the terminals on the converter to the terminals on the batteries**, making sure that they are both polarised, i.e. + to +, - to -. The electrical appliances you wish to power are then simply plugged into the 220/230 V socket(s) on the inverter (note: the sum of the power ratings of the appliances plugged into the inverter must be less than the power rating indicated for the inverter). It is often a good idea to use this converter to supply a small electrical panel secured by fuses (and/or circuit breakers) from which several electrical lines will supply the consumers.

There are two types of inverter: quasi-sine and pure-sine. The electrical signal emitted by a quasi-sine inverter is less regular than that of a pure-sine inverter. This means that **the use of a quasi-sine inverter is recommended with electrical appliances that are neither inductive nor electronic**: incandescent lighting, irons, coffee makers, hotplates, ovens, convectors,   fridge,

radio,   TV   TV, etc.

For other equipment (plasma or LCD screens, computers, measuring equipment, etc.) we strongly recommend the use of pure sine converters.

The **distance between the Regulator - Battery and Battery - Inverter cables** should be **kept to a minimum** (less than 6m) to avoid electricity losses. The vast majority of manufacturers offer equipment that is accurate to within 1 to 2m.

CHAPTER 5

Other sources of renewable energy

Supplied by the sun, wind, the heat of the earth, waterfalls, seaweed or the growth of plants, renewable energies generate little or no waste or polluting emissions. They help to combat the greenhouse effect and CO2 emissions into the atmosphere, facilitate the rational management of local resources and create jobs.

Chapter Overview : Other sources of renewable energy

5.1 Renewable energy families

5.1.1 Solar energy

5.1.2 Wind energy

5.1.3 Hydropower

5.1.4 Biomass5.1.5 Geothermal energy

5.2 The different types of renewable energy in the world

5.3 Profitability

5.1 Renewable energy sources [8]

Solar energy (photovoltaic solar, thermal solar), hydroelectricity, wind power, biomass and geothermal energy are all inexhaustible sources of energy, compared with "stock energy" derived from fossil fuel deposits that are in short supply: oil, coal, brown coal and natural gas. Enter the world of renewable energies: Which energy sources? For what needs? How can they be harnessed and transformed? In what form can they be used?

- Solar energy

> Photovoltaic solar energy

> Low-temperature solar thermal energy

> High-temperature solar thermal energy

- Wind energy

- Hydropower - Hydroelectricity

> Large-scale hydraulics

> Small-scale hydropower

> Marine energies

- Biomass

> Wood energy

> Biogas

> **Biofuels**

- **Geothermal energy**
- **Bioclimatic architecture**

5.1.1 Solar energy

> Low-temperature solar thermal energy

The sun's rays, captured by glass thermal collectors, transmit their energy to metal absorbers - which heat a network of copper pipes through which a heat transfer fluid circulates. This heat exchanger in turn heats the water stored in a hot water tank. A solar water heater produces domestic hot water or heating, generally distributed by a "direct solar floor".

All devices that act as solar thermal collectors are increasingly integrated into bioclimatic architecture projects (solar houses, greenhouses, collector walls, Trombe walls, etc.).

> High-temperature solar thermal energy

By concentrating solar radiation on a collector surface, very high temperatures can be obtained, generally between 400 C and 1,000 C.

Solar heat produces steam, which powers a turbine, which in turn powers a generator that produces electricity - heliothermodynamics.

Three distinct technologies are used in concentrated solar power plants:

- In parabolic concentrators, the sun's rays converge on a single point, the focus of a parabola.
- In tower power plants, hundreds or even thousands of mirrors (heliostats) follow the path of the sun and concentrate its rays on a central receiver at the top of a tower.
- Third technology: parabolic trough collectors concentrate the sun's rays onto a heat-transfer tube located at the focus of the solar collector. After several years of dormancy, the high-temperature solar industry is making a strong comeback, particularly in the sun belt countries.

Figure 5.1 : Solar thermal energy

5.1.2 Wind energy

Wind energy is the energy of the wind, whose driving force (kinetic energy) is used to propel sailing boats and other vehicles, or transformed by means of an aerogenerator, such as a wind turbine or a windmill, into energy that can be used in a variety of ways. Wind energy is an energy renewable energy.

Wind power is an intermittentsource energy intermittente intermittent It is not produced on demand, but according to meteorological conditions; it therefore requires storage facilities or replacement production during its periods of unavailability. Wind power generation can be predicted with a fair degree of accuracy. Its share of global electricity production was 4.8% in 2018 and is estimated at 5.3% in 2019. The main producing countries are China (28.4% of the world total in 2019), the United States (21.2%) and Germany (8.8%).

5.1.3 Hydropower

Like the watermills of yesteryear, hydroelectricity or the production of electricity by harnessing water first appeared in the middle of the 19th century. The water turns a turbine that drives an electric generator that injects kilowatt-hours into the grid.

Hydropower accounts for 19% of total electricity production worldwide. It is the most widely used renewable energy source. However, not all the world's hydroelectric potential has yet been exploited.

> Large-scale hydraulics

This is the energy produced by dams. This energy is made up of two types: large-scale hydropower and small-scale hydropower. The difference between these two names is, on the one hand, related to the electrical power and, on the other hand, depends on the threshold set by the European Commission. Hydroelectricity, or the production of electricity by harnessing water, first appeared in the middle of the 19th century.

The water turns a turbine which drives an electric generator that injects the kilowatt-hours into the grid.

Hydraulic energy has been used for centuries to produce mechanical power. Hydroelectricity began to develop in the 1880s (invention of the turbine in France in 1827).

By the end of the 19th century, electric turbines had almost completely replaced mechanical power generation in Europe. The development of grids and the quest for economies of scale led to the development of large-scale hydroelectric power in the 1930s, to the detriment of small-scale installations.

Figure 5.2: Large hydropower

> Small-scale hydropower

If all small power plants are grouped together under the term small hydropower plant (SHPP), a distinction is made between the pico-power plant: less than 20 kW, the micro-power plant: from 20 kW to 500 kW, the mini-power plant: from 500 kW to 2 MW, and the small power plant: from 2 to 10 MW.

Built on a run-of-river basis, small-scale hydroelectricity does not require impoundment or occasional emptying that could disrupt the hydrology, biology or quality of the water.

Micro hydroelectric power stations operate in the same way as large dam power stations that harness the energy of rivers. France's potential for the creation of PCHs is estimated at at least 1,000 MW.

As a decentralised energy source, small-scale hydroelectricity maintains or creates economic activity in rural areas.

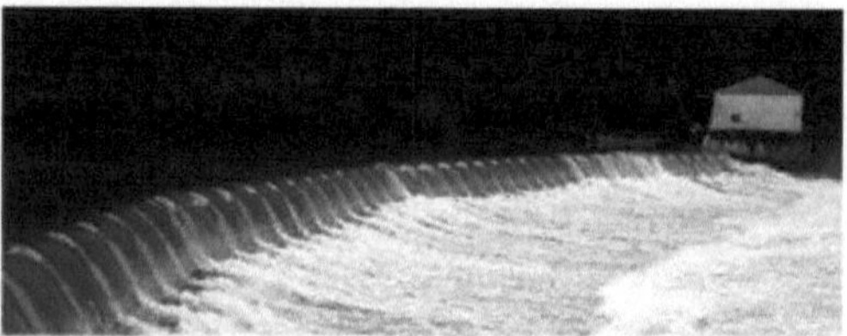

Figure 5.3: Small-scale hydropower

> Marine energies

The marine energy sector, also known as ocean energy or thalasso energy, involves developing technologies and harnessing the natural energy flows provided by the seas and oceans. These include: swell, wave energy, current energy, tidal energy and ocean thermal energy (OTE), which works on the thermal gradient between the layers of water at the surface and at depth.

Marine hydroelectricity uses techniques that are well known, such as the tidal power plant at La Rance (tidal barrage), or that are currently being tested: wave generators (oscillating water column systems, tidal surge systems), tidal turbines (underwater propellers or underwater wind turbines), flat flapping or oscillating wings, floating paddle wheels, etc.

Figure 5.4: Marine energy

5.1.4 Biomass

It comprises three main families:

- Wood energy or solid biomass
- Biogas
- Biofuels

These are all materials of biological origin used as fuels for the production of heat, electricity or fuels.

- **Wood energy or solid biomass**

Wood is a renewable energy source. It is the main wood resource, but other organic materials such as straw, solid crop residues, maize bunches, bagasse from sugar cane and olive pomace must also be taken into account.

In France, as in most other European countries, forest harvesting remains lower than the natural growth of the forest, so the carbon balance is positive.

Today, there are innovative and efficient wood-fired appliances available to individuals, local authorities and industry alike. Biomass boilers burn different types of biofuel: wood pellets, logs, forestry chips, sawdust or shavings.

Figure 5.5: Solid biomass

- **Biogas**

Biogas is released when organic matter decomposes through a fermentation process (methanisation). It is also known as "renewable natural gas" or "marsh gas", as opposed to gas of fossil origin.

A mixture of methane and carbon dioxide plus a few other components, biogas is a combustible gas.

It is used to produce heat, electricity or biofuel.

Biogas can be collected directly from landfill sites or produced in methanisation units.

By-products from the agri-food industry, sludge from sewage treatment plants, slurry, animal or agricultural waste can be methanised in industrial units.

Figure 5.6: Biogas

- Biofuels

Biofuels, sometimes called agrofuels, are derived from biomass. There are two main industrial sectors: ethanol and biodiesel. They can be used pure, as in Brazil (ethanol) or Germany (biodiesel), or as additives to conventional fuels.

In France, 70% of ethanol is produced from sugar beet and 30% from cereals. Biodiesel is produced from oilseeds (rapeseed, sunflower).

Geothermal energy is the exploitation of heat stored underground. There are two main uses for geothermal resources: electricity generation and heat production. Depending on the resource, the technique used and the requirements, there are many different applications.

Figure 5.7: Biofuel

5.1.5 Geothermal energy

The guiding criterion for defining the sector is temperature. Geothermal energy is classified as "high energy" (over 150°C), "medium energy" (90 to 150°C), "low energy" (30 to 90°C) and "very low energy" (under 30°C).

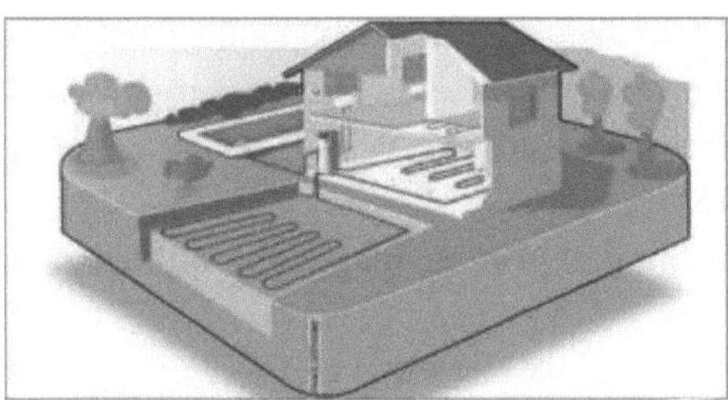
Figure 5.8: Geothermal energy

Bioclimatic architecture

Passive architecture, solar houses, positive energy buildings, high environmental quality, high energy performance ... are just some of the names used to describe bioclimatic architecture.

This method of architectural design involves finding the best balance between the building, the surrounding climate and the comfort of the occupant.

Bioclimatic architecture makes the most of the sun's rays and the natural circulation of air to reduce energy needs, maintain pleasant temperatures, control humidity and encourage natural lighting.

Figure 5.9: Bioclimatic architecture

5.2 The different types of renewable energy in the world

Renewable resources are varied and inexhaustible. Their conversion into thermal, chemical or electrical energy poses few human or ecological dangers. What's more, production can be centralised or decentralised. On the other hand, it is characterised by relatively low efficiency, high cost and intermittence of the resource. Systems using solar, wind and hydro power, as well as biomass, are in operation in many parts of the world. They are becoming increasingly efficient and profitable.

However, the use of renewable resources, apart from large hydroelectric plants, is generally limited to isolated sites where the cost of renewable systems becomes competitive with other means of electricity production because of the very high cost of electricity transmission. [1]

5.3 Profitability [16]

Solar and wind energy will provide more than a third of the world's energy by 2040.

But today, there is hope for the transition to clean energy. According to a report by Bloomberg (the world's largest financial news agency), solar and wind power will provide more than a third of the world's energy by 2040. The report goes further and predicts that **clean energy will soon be more profitable than coal or oil.**

Indeed, the spectacular falls in the production costs of solar energy suggest that in five years' time, alternative energies will be cheaper than coal in many countries, including China, the UK and India. **France should also follow this favourable trend to some extent.**

Bloomberg data adds that the cost of electricity produced by photovoltaic panels has fallen by almost 75% since 2009 and is expected to fall by 66% by 2040. The cost of producing wind energy on land has fallen by 30% since 2009, and this trend is set to continue between now and 2040. A much better idea of the situation can be gained from the report by the Commission de Regulation de l'Energie (CRE) on the profitability of renewable energies. This report reveals a clear upturn in the profitability of new energies. All this suggests that **the energy transition will soon be surfing the waves of profitability**. Investing in renewable energies is now a viable alternative savings solution. Irena experts estimate that every time the world's **renewable energies** double in size, costs fall by 14% for offshore wind power, 21% for onshore wind power, 30% for concentrated solar power and 35% for photovoltaics.

BIBLIOGRAPHY

[1] : Etude D'un Systeme Autonome De Production D'energie Couplant Un Champ Photovoltaique, Un Electrolyseur Et Une Pile A Combustible : Rëalisation D'un Banc D'essai Et Modëlisation, thesis, Docteur de l'Ecole des Mines de Paris, Dec.2003.

[2]

energy-system

[3] : ftp://ftp.fsr. ac.ma/cours/physique/bargach/chap12.pdf

[4]   https://fr.wikipedia.org/wiki/%C3%89nergie renewable

[5] : https://fr.wikipedia.org/wiki/Combustible nucl%C3%A9aire

[6] : https://fr.wikipedia.org/wiki/%C3%89nergie%C3%A9olienne

[7] : https://eolienne.f4jr.org/parc wind

[8] : Cours decouverte systemes energetiques autonomes , M2 ESE (power point version), Mme MAZOUZ-LEKHAL.N, Dec. 2017.

[9] : https://www.quelleenergie.fr/economies-energie/eolienne-domestique/

[10]N. KHORCHEF: "Etude du convertisseur Superboost applique aux systemes photovoltaique", Faculte Genie Electrique, Universite des Sciences et de la Technologie d'Oran Mohamed Boudiaf USTOMB, These de Magistere 2009.

[11]N. MAZOUZ: "developpement d'un convertisseur DC/DC pour l'optimisation du rendement d'un systeme photovoltaique", Faculte Genie Electrique, Universite des Sciences et de la Technologie d'Oran Mohamed Boudiaf USTOMB, These de Doctorat ES-Science 2014.

[12]N. MAZOUZ: "Controle flou d'un generateur photovoltaique (GPV) alimentant un systeme moteur pompe", Faculte Genie Electrique, Universite des Sciences et de la Technologie d'Oran Mohamed Boudiaf USTOMB, These de Magistere 2005.

[13]www.hespul.org, Report written by Violaine Didiers under the direction of Bruno Gaiddon.

[14]BENATIALLAH.D : "Determination of the solar deposit by satellite imagery with integration in a geographic information system for the south of Algeria" faculte des sciences et de la technologie, universite africaine Ahmed draia Adrar, These de Doctorat ES- Science 2019.

[15]:   https://www.ecolodis-solaire.com/conseils/presentation-generale-d-une-installation-photovoltaics-for-isolated-site-presentation-of-a-typical-photovoltaic-installation-29

[16]: https://blog.tudigo.co/investir-dans-les-energies-renouvelables-un-pari-rentable/

Printed by Books on Demand GmbH, Norderstedt / Germany